SpringerBriefs in Biology

For further volumes:
http://www.springer.com/series/10121

Teiji Sota • Hideki Kagata • Yoshino Ando
Shunsuke Utsumi • Takashi Osono

Species Diversity and Community Structure

Novel Patterns and Processes in Plants, Insects, and Fungi

 Springer

Teiji Sota
Department of Zoology
Graduate School of Science
Kyoto University
Kitashirakawa-Oiwake-cho, Sakyo-ku
Kyoto 606-8502, Japan

Yoshino Ando
Center for Ecological Research
Kyoto University
2-509-3 Hirano, Otsu
Shiga 520-2113, Japan

Takashi Osono
Center for Ecological Research
Kyoto University
2-509-3 Hirano, Otsu
Shiga 520-2113, Japan

Hideki Kagata
Center for Ecological Research
Kyoto University
2-509-3 Hirano, Otsu
Shiga 520-2113, Japan

Shunsuke Utsumi
Field Science Center for Northern Biosphere
Hokkaido University
Moshiri, Horokanai
Hokkaido 074-0741, Japan

ISSN 2192-2179   ISSN 2192-2187 (electronic)
ISBN 978-4-431-54260-5   ISBN 978-4-431-54261-2 (eBook)
DOI 10.1007/978-4-431-54261-2
Springer Tokyo Heidelberg New York Dordrecht London

Library of Congress Control Number: 2013948397

Printed on acid-free paper

Springer is part of Springer Science+Business Media (www.springer.com)

Foreword

While the twentieth century was the century when researchers tried to discover "the general basic principles of organisms," the twenty-first century is expected to be the century when researchers try to understand "the evolution and diversity of organisms" on the basis of such general principles of organisms by integrating various disciplines such as morphology, physiology, and ecology.

The chief difficulty in studying "the evolution and diversity of organisms" lies in the fact that we have to consider factors at various levels ranging from the genome to the ecosystem. As taking various factors into account may cause a loss of focus, traditional studies have been restricted to analyzing only one individual level or factor. However, unfortunately, the current research and education system based on such a compartmentalized approach is inadequate for incisively studying "the evolution and diversity of organisms."

In order to solve these problems, we should strongly emphasize the necessity for joint studies and integration of the education programs between micro-level biology (genomic science, evolutionary developmental biology, genetic science, cell biology, neurobiology, molecular physiology, and molecular evolutionary studies) and macro-level biology (primatology, anthropology, ethology, environmental biology, evolutionary taxonomy, and so on) to young biologists. We launched a new education program in Kyoto University, called "Global COE program for Evolution and Biodiversity Research" to promote such integrative studies at various levels, and have succeeded in initiating novel currents of study of biodiversity that led rather than followed those in the rest of the world. To this aim, we decided to publish six books in "SpringerBriefs in Biology" which we hope will stimulate interest in such novel approaches on the evolution and diversity of organisms in the world and among young biologists.

In this book, we describe recent findings about how metagenomics studies are revealing the previously unexplored hyperdiversity of tropical fungi, how herbivore–plant interactions affect community/ecosystem genetics, and how diversification and speciation occur in insects.

First, we provide a literature review that not only includes patterns of fungal diversity emerging from metagenomics but also summarizes methodologies that will be of interest to readers who are studying mycology and ecology and are interested in this developing research field of biodiversity science. Previous studies of fungal metagenomics using next-generation sequencing (NGS) techniques have been carried out in temperate regions of Europe and North America. We describe our research project which is unique in that it was performed in Asian tropical forests that harbor an unexplored diversity of fungi. Canopy crane facilities were used to access the canopy of tropical rainforests at heights of 30–50 m, and the data obtained provide basic information useful for understanding the hyperdiversity of tropical fungi.

Genetically controlled plant traits are well established as a fundamental determinant of population, community, and ecosystem properties in terrestrial ecosystems. For example, plant traits play an important role in determining above- and belowground community structure, litter decomposition processes, and nutrient cycling (i.e., community/ecosystem genetics). It is important to recognize that herbivores also play an important role in determining such properties through direct and indirect interactions. Here, we review what is known about community/ecosystem genetics and highlight how herbivores are important to modify the community and ecosystem consequences of genetically controlled plant traits.

Insects account for more than half of the known species of organisms on our planet, and the process of their diversification is a central issue of evolutionary biology. Focusing on divergences in beetles' ability to fly and in the reproductive season of a moth group, we show that allopatric and allochronic speciation in insects can be promoted by divergence in life-history strategy.

Kiyokazu Agata
Professor, Department of Biophysics, Kyoto University
Project Leader of Kyoto University Global COE program
"Evolution and Biodiversity"

Preface

Understanding the present status of biodiversity and the evolutionary and ecological processes creating enormous biodiversity on our planet is acutely important to contemporary biologists in the era of the biodiversity crisis caused by increasing human impacts on natural environments. This book, *Species Diversity and Community Structure: Novel Patterns and Processes in Plants, Insects, and Fungi*, deals with three different approaches to contemporary issues in biodiversity research conducted by three groups at Kyoto University. Because of the vast species richness and the extreme complexity of biological interactions, biodiversity research includes diverse subjects and methodologies.

First, revealing the precise species richness of different taxonomic groups, and ultimately the total number of species on our planet at present, is an important objective of biodiversity research. In this decade, DNA barcoding has become a powerful tool to unveil species richness in various groups of organisms in lieu of classical taxonomy based largely on morphological evidence. More recently, the advances in high-throughput sequencing technologies have further accelerated the speed of species inventory, especially for invisible microorganisms. Takashi Osono and his colleagues at the Center for Ecological Research study species diversity and functioning of fungi in Asian regions using metagenomic approaches. In Chap. 1, Osono reviews the current status and methodological aspects of the fungal diversity research, especially the metagenomic approach using 454 pyrosequencing. He discusses potentials and problems in the metagenomic studies of fungi.

Complex ecological and evolutionary interactions between plants and herbivorous insects can affect community structures and ecosystem functioning. Recent studies in community and ecosystem genetics shed light on these complex interactions with novel approaches incorporating genetic perspectives. Hideki Kagata and his colleagues at the Center for Ecological Research have conducted experimental studies to reveal the interrelationship among herbivore traits, plant traits, and arthropod community structure, considering genetic variation and phenotypic plasticity in plant defenses against herbivores. In Chap. 2, Hideki Kagata, Yoshino Ando, and

Shunsuke Utsumi review various aspects of the study of complex interactions between herbivorous insects and plants in the community and ecosystem contexts.

Lastly, the speciation process is a central topic in the study of biodiversity, as species diversity on our planet cannot increase without speciation. My colleagues and I in the Department of Zoology have studied various aspects of speciation in insects related to adaptation through natural and sexual selection. Here I introduce two of our recent studies on the speciation processes, which are related to particular life history evolutions. First, the hypothesis that evolution of flightlessness in winged insects can promote allopatric speciation is examined in beetles, one of the most species-rich insect groups. Second, I demonstrate that climatic disruption of reproductive period leads to allochronic speciation in a winter moth group and suggest that temporal isolation provided by periodical climatic harshness in a seasonal environment may have promoted diversification of insects. These aspects of speciation through particular life history evolution may be important in diversification of insect species in heterogeneous environments in space and time.

Kyoto, Japan   *Teiji Sota*
                                                              on behalf of all the authors

Contents

Chapter 1
Metagenomic Approach Yields Insights into Fungal Diversity and Functioning

Abstract Recent advances in molecular biological methods have drastically changed our understanding of the diversity, phylogeny, ecology, and evolution of fungi. The purpose of this chapter is to address recent progress in our knowledge about the biodiversity of fungi, with emphasis on the potential impact of pyrosequencing to estimate fungal diversity in environmental samples. Progress in DNA sequence-based techniques notably enables us not only to overcome potential flaws of traditional mycological techniques but also to evaluate fungal richness more efficiently and reliably. Especially, the development of next-generation sequencing technologies has revolutionized large-scale sequencing of environmental fungal DNA. A number of papers have been published regarding metagenomic analysis of fungal diversity in environmental samples, using Roche 454 pyrosequencing since 2009. The number of publications on fungal metagenomics will be accelerated, but there are potential methodological difficulties that have not been solved yet. DNA barcoding with universal genetic markers and high-quality sequence databases becomes more important as reliable taxonomic affiliations of numerous MOTUs are necessitated. Further studies are needed to test the applicability of pyrosequencing to such hot spots of fungal diversity as tropical forests and to explore functional aspects of fungal populations.

Keywords Biodiversity • Fungi • Metagenomics • Next-generation sequencing • Pyrosequencing

1.1 Introduction

Recent advances in molecular biological methods have drastically changed our understanding of the diversity, phylogeny, ecology, and evolution of fungi. The purpose of this chapter is to address recent progress in our knowledge about the biodiversity of fungi, with emphasis on the potential impact of pyrosequencing to estimate

T. Sota et al., *Species Diversity and Community Structure: Novel Patterns and Processes in Plants, Insects, and Fungi*, SpringerBriefs in Biology, DOI 10.1007/978-4-431-54261-2_1, © The Author(s) 2014

fungal diversity in environmental samples. This metagenomic approach will provide light to guide future explorations of yet-to-be-discovered fungal diversity in biodiversity "hot spots," such as the canopy of tropical forests, and may be useful for investigating functional aspects of fungal diversity.

After briefly introducing fungal diversity, I summarize methods to study fungal diversity and our current understanding of the global fungal inventory to highlight the potential applicability of pyrosequencing to studying the biodiversity of fungi. Then I synthesize papers on fungal metagenomics published since 2009 to show the broad usefulness of pyrosequencing and DNA barcoding for the study of fungal diversity in environmental samples and address methodological difficulties yet to be solved. Finally, I discuss future directions related to the application of pyrosequencing for assessment of tropical fungal diversity and of the functioning of fungal populations.

1.2   An Overview of Fungal Diversity and Functioning

Fungi are heterotrophic, eukaryotic microorganisms comprising molds, mushrooms, yeasts, and lichens. They have remarkably diverse life histories that make indispensable contributions to the maintenance of ecosystems and human life (Stajich et al. 2009). According to the latest Dictionary of the Fungi (Kirk et al. 2008), fungi are defined as those organisms without plastids, having absorptive nutrition (osmotrophic), never being phagotrophic, and lacking an amoeboid pseudopodial phase. Fungal cells are microscopic and are mostly either hyphae that grow at their tips and branch to produce multicellular mycelia, or unicellular yeast that duplicates by budding (Jennings and Lysek 1999). The essential nutritional mode of fungal cells is that they excrete a suite of extracellular enzymes that break down organic molecules in the environment and thereby enable the absorption of substrates through the cell wall for intracellular metabolism.

Fungi are diverse in terms of their phylogeny and taxonomy. The origin of fungi is estimated to date back to approximately a billion years ago, give or take 500 million years (Taylor and Berbee 2006). Ancestral fungi, possibly aquatic, unicellular eukaryotes making sporangia containing zoospores (Cavalier-Smith 1987), have now diversified into several major branches, which are divided into two major groups, basal fungi and dikarya, according to the Fungal Tree of Life project (James et al. 2006; Hibbett et al. 2007). The basal fungi have several major clades, including Chytridiomycota, Blastocladiomycota, Neocallimastigomycota, which were formerly classified into the traditional Chytridiomycetes, and Mucoromycotina, Kickxellomycotina, Zoopagomycotina, Entomophthoromycotina, and Glomeromycota, which were formerly classified into the traditional Zygomycota. The Dikarya is a subkingdom embracing the two largest fungal phyla, Ascomycota and Basidiomycota, which typically have cells with two nuclei (dikaryons).

Fungi are diverse in their functioning, as interaction with other life is crucial for the heterotrophic lifestyle. Fungal hyphae are an adaptive form that allows extension on the surface and penetration of the internal tissues of other larger organisms, such as plants, animals, and fungi themselves. Yeasts are suitable for dispersal and reproduction in the liquid phase, such as the sap of plants and body fluids of animals. Those fungi that live on living organisms are called parasites. Some parasitic fungi, such as mycorrhizal and endophytic fungi, are regarded as mutualistic symbionts as they provide net benefits to their hosts at the expense of costs of photosynthetic assimilates supplied to the fungal parasites. Parasites that are harmful to their hosts, such as rusts and smuts, are known as pathogens. In contrast, fungi are called saprobic when they live on dead organisms. Saprobic fungi have become equipped during their evolutionary history with an array of extracellular enzymes, such as ligninases, some of which are unique to the kingdom Fungi (Zamocky and Obinger 2010) and play crucial roles in the turnover of organic matter in soils (Sinsabaugh 2010). This supports the general notion that fungi are primarily effective decomposers in ecosystems (Boddy et al. 2008). Lichens are composite organisms in a symbiotic association between a filamentous fungus (mostly in the Ascomycota) and a microalga and/or a cyanobacterum, and are unique in that they are primary producers in terms of their functioning in ecosystems (Lutzoni and Miadlikowska 2009).

Because of their unique functions, fungi are an indispensable component of ecosystem services for human well-being. For example, saprobic activities of decomposer fungi ensure the transformation and accumulation of soil organic matter and the recycling of essential nutrients in plant-soil systems, thus contributing to such regulating and supporting services as primary production, carbon sequestration and decomposition in soils, and purification of water through soil humus. Fungi account for provisional services through the production of edible mushrooms and alcohol as foods and drinks, the activity of transforming and detoxifying pollutants, and the ability to produce pharmaceuticals and biochemicals as secondary metabolites. Regarding the last of these abilities, bioactive natural products from endophytic fungi (latent invaders of healthy-looking plant tissues) are attracting particular attention (Aly et al. 2010). Fairy rings, naturally occurring rings or arcs of mushrooms, are a good example of cultural services that fungi provide, as they give spiritual inspiration and play a prominent role in European folklore as places where elves dance. Hence, fungi in natural ecosystems are diverse in their benefits to and intimate associations with human beings.

1.3   Methods to Study Fungal Diversity

Several approaches have been proposed and adopted to study species richness of fungi in various environments and are broadly divided into mycological and molecular biological techniques (Fig. 1.1). These approaches have both usability and limitations that have been documented repeatedly in the literature (Gams 1992; Hibbett et al. 2011; Su et al. 2012) and are summarized here briefly.

Traditional mycological techniques to delimit, describe, identify, and enumerate fungal species involve direct morphological observation of macroscopic and microscopic structures of reproductive organs, such as fruiting bodies of macrofungi.

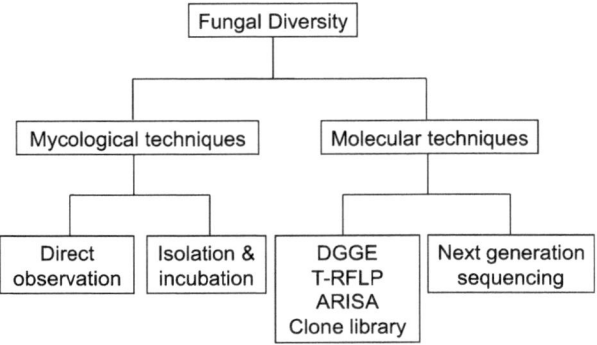

Fig. 1.1 Representative methods to study fungal diversity

Microfungi present in soils and animal and plant tissues (e.g. leaves, stems, and roots) are isolated from these substrata first, and then incubated on nutrient media to induce the formation of reproductive structures, such as conidia and conidiophores, before the identification. Morphological observations are essential for taxonomic species identification but require knowledge and professional experience regarding fungal taxonomy. Major flaws of mycological techniques include difficulties in detecting cryptic species and uncultivable species, and the selectivity of nutrition and culture conditions during the isolation and incubation.

Molecular biological techniques involve the measurement of biochemical molecules representative of fungal taxa (i.e. biomarkers), such as amino acid composition and deoxyribonucleic acid (DNA). Progress in DNA sequence-based techniques now notably enables us not only to overcome potential flaws of mycological techniques but also to evaluate fungal richness more efficiently and reliably. Common DNA sequence-based approaches to the study of fungal diversity are denaturing gradient gel electrophoresis (DGGE), terminal restriction fragment length polymorphism (T-RFLP), automated ribosomal intergenic spacer analysis (ARISA), and clone library methods (Okubo and Sugiyama 2009; Goel et al. 2011). More recently, the development of next-generation sequencing (NGS) technologies has revolutionized large-scale sequencing of environmental microbial DNA. NGS employs newer methods than the automated Sanger method, which is referred to as a "first-generation" technology. A suite of NGS platforms is currently available (Glenn 2011), including Applied Biosystems SOLiD, Illumina Sequencing, and 454 Life Science/Roche massively parallel pyrosequencing. The common primary advantage of NGS over the "first-generation" method is the inexpensive and rapid production of large volumes of sequence data. That is, NGS has drastically increased the number of bases obtained per sequencing run and decreased the cost per base (Metzker 2010). NGS is multipurpose and has been used for genome analysis (Voigt and Kirk 2011), isolation of microsatellite markers (Malausa et al. 2011), and metagenomics. Especially, NGS has greatly facilitated metagenomic studies of microbes in environments, first for prokaryotes and recently also for eukaryotes (Nowrousian 2010). Roche 454 pyrosequencing is the only platform yielding DNA templates [about 250 base pair (bp)] long enough to be considered for rigid use in a species-level barcoding framework and suitable for fungal metagenomics of environmental samples. The application and usefulness of 454 pyrosequencing for fungal metagenomics was briefly summarized in Hibbett et al. (2011) and will be reviewed in more detail in the next section of this chapter.

1.4   How Many Fungal Species on the Globe?

A number of estimates have been proposed for the magnitude of fungal diversity in the world. Hawksworth (1991) is a landmark paper that proposed the commonly referred to figure of 1.5 million (M) estimated fungal species. This figure was based primarily on the observed ratio between the species richness of vascular plants and fungi in countries where fungal inventories had been sufficiently well established (e.g. one plant to six fungal species in the British Isles). The local plant-to-fungi ratios were then extrapolated to a conservative number of known vascular plant species worldwide to give the 1.5 M estimate of fungal richness. A decade later, Hawksworth (2001) revisited this and other 14 global richness estimates (ranging from 0.5 to 9.9 M) published since 1990 and argued that it is prudent to retain 1.5 M as the current working hypothesis for the number of fungi on the earth. Hawksworth (1991, 2001) discussed limitations and potential critiques of this "diversity ratio" approach and suggested the need for further inventories at "hot spots" of fungal diversity, such as live leaves on the canopy of tropical forests, which are believed to harbor potential hyper-diversity of endophytic fungi (Blackwell 2011).

Using a similar "diversity ratio" approach, Schmit and Mueller (2007) estimated the "lower limit of global fungal diversity." This study combined individual estimates of global species richness for major fungal groups that are better known in terms of the inventory and geographic distribution (e.g. macrofungi, lichens, and aquatic fungi) and took regional variations (e.g. temperate versus tropical area) of plant-to-fungi ratios for these groups into account. Using conservative assumptions at all stages of analyses, Schmit and Mueller (2007) reached 0.71 M as a minimum number of fungal species.

The most recent study (Mora et al. 2011) used a different method to estimate the global number of eukaryotic species and predicted the species richness of fungi to be 0.61 M and that of all eukaryotes to be 8.7 M. Data of "valid" species available in the database Catalogue of Life (as of 2006) were used in this analysis. Firstly, asymptotic parametric regression models were applied to the temporal accumulation curve (from AD 1750 to 2006) of each taxonomic rank (phylum, class, order, family, and genus) for major taxonomic groups (e.g. Animalia, Plantae, Fungi) to predict the asymptotic number of these taxa. Then, these asymptotic numbers were regressed with the hyper-exponential function against their numerical ranks from one (phylum) to five (genus), and the resulting regression model was used to extrapolate to the species level (rank = 6) to obtain the estimated number of species for fungi and other taxonomic groups. It is unclear whether the estimate (0.61 M) for fungi was an underestimate or not, but it appears to be preliminary for two reasons. First, this calculation was based on data of 43,000 species that was rather limited, as currently 97,861 (about 0.1 M) fungal species are described (Kirk et al. 2008). Secondly, the fits of asymptotic parametric regression models for fungal family and genus were low, probably because many families and genera are yet to be described. This made the predictions of the asymptotic number of these taxa, and the extrapolated number of fungal species as well, unreliable. Nevertheless, Mora et al. (2011) is useful as it presented an alternative approach for the estimation of the global number of fungal species.

How can recent advances in the analysis of environmental DNA contribute to the global richness estimate of fungi? O'Brien et al. (2005) performed a clone library method to evaluate soil fungal diversity in two temperate forests. A total of 863 fungal ITS sequences were obtained and classified into 412 sequence types (threshold at 97 % similarity), which included unidentifiable sequence types at high rates. Then, the ratios of ITS-based estimates of soil fungal richness to vascular plant richness of the study sites were extrapolated to worldwide vascular plant species richness. This yielded tentative estimates of global soil fungal species richness ranging from 3.5 to 5.1 M, which falls within the range summarized in Hawksworth (2001).

These estimated numbers of global fungal species richness (0.5–9.9 M) go far above the number of fungal species catalogued so far (0.1 M, Kirk et al. 2008). Using the number of described species and the current rate of description of new species (1,196 species per year averaged for the last 10 years), Hibbett et al. (2011) noted that it would take 1,170–4,170 years to describe 1.5–5.1 M fungal species. Obviously, the pace of species description should be accelerated dramatically to complete the global fungal inventory. This calculation led Hibbett et al. (2011) to argue the need for formal naming of environmental fungal DNA sequences and possible options for sequence-based classification systems of fungi, as the database of molecular operational taxonomic units (MOTUs) obtained from environmental samples with NGS is anticipated to expand soon (see the next section).

1.5   Fungal Metagenomics and Methodological Considerations

A total of 24 papers have been published regarding metagenomic analysis of fungal diversity in environmental samples, using 454 pyrosequencing (as of January 10, 2012; Table 1.1). The samples analyzed included plant roots and soils, as well as plant leaves, indoor dust, oral rinse of humans, animal feces, and lake water. Most of the publications came from temperate regions, while two studies were carried out in tropical regions (Tedersoo et al. 2010; Arfi et al. 2012) and two studies comprised intercontinental comparison (Amend et al. 2010b; Moora et al. 2011). The genetic markers used were ITS1 (11 cases), ITS2 (six cases), SSU or 18S (nine cases), and large subunit (LSU) or 28S (four cases) regions of nuclear ribosomal DNA (nrDNA) (Fig. 1.2). As many as 300,000 DNA sequences per study and 50,000 per sample were analyzed for fungal MOTU richness in these studies.

Table 1.1 Studies of metagenomic analysis of fungal diversity in various environmental samples

Reference	Taxon or functional group of special interest	Study site	Substratum	PCR primers	Genetic marker	Number of sequences after quality filtering	Threshold %	Number of MOTUs	Number of singleton MOTUs (% total)
Fungi on aerial plant parts									
Jumpponen and Jones (2009)	Phyllosphere fungi	Kansas, USA	Live leaves of *Quercus macrocarpa*	ITS1-ITS2	ITS1	18,020	95	689	360 (52 %)
Jumpponen and Jones (2010)	Phyllosphere fungi	Kansas, USA	Live leaves of *Quercus macrocarpa*	ITS1F-ITS4	ITS2	83,554	95	1,232	491 (40 %)
Arfi et al. (2012)	Fungi on aerial and intertidal parts of mangrove trees	Southern Province, New Caledonia	Leaves, trunks, and barks of *Avicennia marina* and *Rhizophora stylosa*	ITS1F-ITS4	ITS1	62,485	98	2,877[a]	na (54 %)
				ITS1F-ITS4	ITS2	82,097	98	3,506[a]	na
				nu_ssu_0817-nu_ssu_1536	SSU (V5)	33,746	98	1,683[a]	na
				nu_ssu_0817-nu_ssu_1536	SSU (V7)	31,216	98	1,455[a]	na
Fungi on plant roots									
Öpik et al. (2009)	Arbuscular mycorrhizal fungi in Glomeromycota	Koeru, Estonia	Roots of 10 plant species	NS31-AM1	SSU	179,279	97	48[a,b]	4 (8 %)
Lumini et al. (2010)	Arbuscular mycorrhizal fungi in Glomeromycota	Sardinia, Italy	Soils of vineyards, pasture, meadow, and forest	AMV4.5NF-AMDGR	SSU	3,189	97	117[b]	37 (32 %)
				NS31-AMmix	SSU	1,003	97	28[b]	9 (32 %)

(continued)

Table 1.1 (continued)

Reference	Taxon or functional group of special interest	Study site	Substratum	PCR primers	Genetic marker	Number of sequences after quality filtering	Threshold %	Number of MOTUs	Number of singleton MOTUs (% total)
Dumbrell et al. (2011)	Arbuscular mycorrhizal fungi in Glomeromycota	Leeds, UK	Mixed plant roots	NS31-AM1[d]	SSU	108,245	97	70[b]	3 (4 %)
Lekberg et al. (2012)	Arbuscular mycorrhizal fungi in Glomeromycota	Zeeland, Denmark	Roots of *Plantago lanceolata*	glo454-NDL22	LSU	213,323	97	32[b]	0 (0 %)
Moora et al. (2011)	Arbuscular mycorrhizal fungi in Glomeromycota	14 sites in Europe and China	Roots of *Trachycarpus fortunei*	NS31-AM1	SSU	109,884	97	73[a,b]	9 (11 %)
Jumpponen et al. (2010a)	Ectomycorrhizal fungi	Kansas, USA	Ectomycorrhizal roots of *Quercus* spp.	ITS1F-ITS2	ITS1	29,910	95	1,077	612 (57 %)
Tedersoo et al. (2010)	Ectomycorrhizal fungi	Korup, Cameroon	Ectomycorrhizal roots of caesalpinioid legumes	ITS1F-ITS2, ITS5-ITS2	ITS1	44,411	97	312	87 (28 %)
Wallander et al. (2010)	Ectomycorrhizal fungi	Tonnersjoheden, Sweden	Mycelia grown into quartz sand buried in *Picea abies* forest soils	ITS1F-ITS4	ITS2	ca. 18,000	98.5	248[a]	184 (43 %)
Hui et al. (2011)	Ectomycorrhizal fungi	Halvala, Finland	Mycelia grown into quartz sand buried in *Pinus sylvestris* forest soils	ITS1F-ITS2	ITS1	8,194	95	161[a]	na

Gottel et al. (2011)	Root endophytes	Tennessee, USA	Surface sterilized roots of *Populus deltoides*	LROR-LR3	LSU	698–18,154	95	511–1,171[c]	na
	Rhizosphere fungi	Tennessee, USA	Rhizosphere soilis of *Populus deltoides*	LROR-LR3	LSU	134–19,818	95	42–315[c]	na
Soil fungi									
Buée et al. (2009)	Soil fungi	Burgundy, France	Soils from pure plantations of six tree species	ITS1F-ITS2	ITS1	25,700–35,600	97	590–1,000	na
Jumpponen et al. (2010b)	Soil fungi	Kansas, USA	Soils within vertical profiles in tallgrass prairie	ITS5-ITS4	ITS2	14,578	95	1,151	554 (48 %)
Rousk et al. (2010)	Soil fungi	Rothamsted, UK	Soils across a pH gradient in an arable soil	5.8S-ITS1f	ITS1	4,700	97	725	na
Lim et al. (2010)	Soil fungi	Islands in the Yellow Sea of Korea	Forest soils	TFungi18F-TFungi18R	SSU	10,166	97	736	na
Lentendu et al. (2011)	Soil fungi	French Alps, France	Soils of alpine tundra	ITS5-ITS2	ITS1	205,131	98	2,847	820 (29 %)
Mello et al. (2011)	Soil fungi	Cahors, France	Soils of truffle grounds	ITS1F-ITS2	ITS1	8,058	97	900	361 (40 %)
	Soil fungi			ITS3-ITS4	ITS2	5,588	97	885	379 (43 %)
Miscellaneuos									
Amend et al. (2010b)	Indoor fungi	Ten countries on six continents	Indoor dusts	ITS1F-ITS4	ITS2	97,557	97	4,473	na

(continued)

Table 1.1 (continued)

Reference	Taxon or functional group of special interest	Study site	Substratum	PCR primers	Genetic marker	Number of sequences after quality filtering	Threshold %	Number of MOTUs	Number of singleton MOTUs (% total)
Ghammoum et al. (2010)	Oral fungi	Ohio, USA	Oral rinse of Homo sapiens	LROR_F-LR5F	LSU	187,668	na	na	na
				ITS1F-ITS2	ITS1	34,049	98	101	na
Liggenstoffer et al. (2010)	Anaerobic gut fungi (Neocallimastigomycota)	Oklahoma, USA	Fecal samples of 30 herbivore animals	MN100-MNGM2	ITS1	48–49,215	95	4–60[c]	na
Monchy et al. (2011)	Freshwater fungi	Massif Central, France	Lake water from two lakes	nu_SSU_0817-nu_SSU_1196	SSU	8,018–16,672	97	ca. 200–700[c]	na
Lucero et al. (2011)	Unculturable fungi in micropropagated plant callus	New Mexico, USA	Callus from seeds of Atriplex canescens	58A1F-LB4	ITS	na	95	10	na
				ENDOITSF-R	ITS	na	95	8	na

na not available

[a]Singletons were excluded

[b]MOTU in the Glomeromycota

[c]Data of individual samples were indicated

[d]Primers used for semi-nested PCR

Fig. 1.2 Schematic diagram of nuclear ribosomal DNA (nrDNA) of fungi. The nuclear multicopy rDNA repeat codes for several ribosomal RNAs (rRNAs), such as small subunit (SSU), 5.8S, and large subunit (LSU). The tandem repeat includes two spacer regions, the intergenic spacer (IGS) and the internal transcribed spacer (ITS). The *black arrow* indicates a commonly used barcode region spanning ITS1, 5.8S, and ITS2. Modified from Fig. 1.2 of Begerow et al. (2010)

A variety of pipelines and bioinformatic tools were used to check quality and classify fungal DNA sequences into MOTUs, adopting the threshold of 95–98.5 % sequence similarity (Table 1.1). Applying average values of intraspecific ITS variability of Ascomycota and Basidiomycota (97–98 % similarity; see below) and a SSU or LSU variability of Glomeromycota (97 % similarity), each study encountered as many as 100 MOTUs for arbuscular mycorrhizal fungi, 1,000 MOTUs for ectomycorrhizal fungi, 3,000 MOTUs for soil fungi, and 3,500 MOTUs for fungi associated with aerial plant parts (Table 1.1). These MOTU numbers explicitly illustrate the efficiency and usefulness of 454 pyrosequencing as a descriptor of fungal diversity in environmental samples. Hibbett et al. (2009) referred to this situation as: "fungal ecology catches fire."

The number of publications on fungal metagenomics will be accelerated in the near future, but there are potential methodological difficulties that have not been solved yet, such as the threshold level of MOTU similarity, the choice of genetic marker, and the handling of singletons. Firstly, it is difficult to apply a single unifying yet stringent upper limit for intraspecific variability to a mixture of fungal MOTUs in environmental samples. Obviously, the number of MOTUs increases with the level of threshold for sequence similarity. Tedersoo et al. (2010) and Lentendu et al. (2011) showed that the number of MOTUs increased abruptly when the ITS1 barcoding threshold exceeded 97 %. As shown in Table 1.1, the studies on arbuscular mycorrhizal fungi consistently used the 97 % criterion for SSU rDNA of the Glomeromycota. However, the other studies varied in the threshold levels (Table 1.1), making the direct comparison of MOTU richness difficult.

In fact, the degree of intraspecific ITS variability differs widely at the fungal species level and at higher taxonomic levels. Nilsson et al. (2008) quantified the intraspecific ITS variability in all fungi for which sequence data were available through the International Nucleotide Sequence Database (INSD). Analyzing the data of 4,185 species, the intraspecific sequence variability was found to range from 0.0 % (*Filobasidiella neoformans*, $n = 114$) to 24.2 % (*Xylaria hypoxylon*, $n = 13$), with a grand average (± standard deviation) of 2.51 ± 4.57 % for 4,185 species. The intraspecific variability was not clearly correlated with the taxonomic affiliation or nutritional mode of the taxa. Species in the comparatively well-studied Ascomycota (an average of 1.96 % for 2,509 species), Basidiomycota (3.33 %, 1,582 species), and Zygomycota (3.24 %, 60 species) stood out as less variable than those in Chytridiomycota (5.63 %, 11 species) and Glomeromycota (7.46 %, 23 species). The result of Nilsson et al. (2008) implies a necessity to apply different threshold levels to delimitate fungal MOTUs that belong to different taxonomic groups constituting fungal assemblages in environmental samples.

In another study, Hughes et al. (2009) measured the percentage base heterozygosity (sequence divergence) of haplotypes of the ITS region derived from a single fruit body of Basidiomycota to represent genetic variation within the same biological species. The percentage varied from 0.17 to 3.27 % for 100 randomly collected fruit bodies, 97 of the 100 fruit bodies had percentage ≤ 2 %, and 99 had ≤ 3 %. The authors suggested similarity threshold 97–98 % for species in Basidiomycota inhabiting their study site, which was similar to the grand average for Basidiomycota in Nilsson et al. (2008). Still, it is difficult to relate MOTU richness directly to species richness because the threshold of sequence similarity is not consistent across different species (Nilsson et al. 2008).

Secondly, use of different genetic markers leads to noticeable differences in taxonomic composition of fungal communities evaluated with 454 pyrosequencing (Lumini et al. 2010; Mello et al. 2011; Arfi et al. 2012). Lumini et al. (2010) compared arbuscular mycorrhizal fungal diversity using two primer pairs, namely, NS3-AMmix (a mixture of AM1, AM2, and AM3), which gave longer but fewer amplicons, and AMV4.5NF-AMDGR, which gave shorter but more amplicons (Table 1.1). The authors noted that a completely satisfactory primer pair was not yet available and that the use of multiple sets was a good strategy for 454 pyrosequencing of arbuscular mycorrhizal fungi. Mello et al. (2011) used two genetic markers, ITS1F-ITS2 and ITS3-ITS4, for amplification of the ITS1 and ITS2 regions, respectively, and showed that the pair ITS1F-ITS2 was more specific for fungi (i.e., produced fewer non-fungal sequences) and produced a higher number of sequences than the pair ITS3-ITS4 (Table 1.1, Fig. 1.2). The class composition changed depending on the pair, and the relative proportions of Dothidiomycetes and Eurotinomycetes were greater in the ITS2 than in the ITS1 regions. Arfi et al. (2012) used two pairs of primers to amplify the ITS and SSU V5-V7 regions of rDNA and sequenced the amplicon libraries from both their 5′ and 3′ ends to obtain the ITS1 and ITS2 regions of the ITS amplicons and the V5 and V7 regions of the SSU amplicons (Table 1.1). The resulting four datasets differed in several abundant taxonomic orders, partly due to the difference in the data availability of reference databases. Similarly to Lumini et al. (2010), Arfi et al. (2012) suggested that using multiple primers is a way to enhance the quality of the description of the taxonomic composition of fungal communities.

Thirdly, there is another difficulty in the handling of singletons. Singletons are MOTUs consisting of only one sequence read. In any 454 dataset, singletons account for less than a few percent in terms of the number of reads but for more than 50 % in terms of the number of MOTUs (Table 1.1). The overwhelming richness of singletons prevents rarefaction curves from reaching asymptotes and makes the richness estimation less reliable. This particularly holds true for fungal communities in soils and in the phyllosphere especially where fungal hyper-diversity is expected (Jumpponen and Jones 2009, 2010). Some authors regarded singletons as artifacts

Table 1.2 Summary of elements that Nilsson et al. (2011) suggested all next-generation sequencing studies of fungal communities should report in a clear and comprehensive way

Step	Elements to report on
Filtering, denoising, and availability of sequence data	Raw sequence data file
	Filtering and trimming
	Sequence denoising
	Number of discarded and retained sequences
	Sequencing depth
	FASTA file
Analysis and taxonomic assignment of sequence data	Details of the genetic marker used
	Type and specifics of sequence clustering
	Sequence data used for taxonomic annotation
	Specification of the taxonomic reference database
	Specification of the taxonomic annotation procedure
	Handling of singletons and MOTUs with few reads
Post-clustering/taxonomic results	Count of MOTUs recovered
	List of MOTUs recovered
	Taxonomic affiliations
	Proportion of fully identified MOTUs
	Phylum-level distribution

and excluded them from data analysis because they contained a strongly elevated proportion of insertions compared with natural intra- and interspecific variation (see Tedersoo et al. 2010). However, there has been no consensus as to whether, or how many, singletons can be regarded as artifacts and thus biologically meaningless.

A few attempts have been made to compare the results between 454 pyrosequencing and other molecular biological methods (Fig. 1.1). Tedersoo et al. (2010) examined the fungal diversity on ectomycorrhizal root tips in a tropical rainforest with pyrosequencing and the traditional Sanger sequencing. Both methods yielded qualitatively similar results, but there were slight, but significant differences in the taxonomic composition of the fungal community. Ovaskainen et al. (2010) applied both 454 pyrosequencing and DGGE to the fungal community in *Picea abies* logs. More fungal species were successfully identified with 454 (30 sp.) than with DGGE (9 sp.), with eight species being common to the two methods. To the knowledge of the author, no studies have compared the fungal community between 454 pyrosequencing and mycological incubation-isolation methods (Fig. 1.1).

Nilsson et al. (2011) argued a need for a standardized procedure to describe and publish NGS datasets of fungal communities and proposed elements that should be specifically reported so as to deliver data and results in a clear and comprehensive way (Table 1.2). This would make the data and results available to the scientific community and would be useful for further comparisons of datasets from different empirical studies (e.g. Unterseher et al. 2011). Unterseher et al. (2011) compared the patterns of diversity of phyllosphere, ectomycorrhizal, and arbuscular mycorrhizal fungi examined with pyrosequencing and published in different papers The rarefaction curve for arbuscular mycorrhizal fungi displayed a long and stable plateau, and four models of richness estimators predicted that the MOTU richness was saturated at about 10,000 reads. In contrast, phyllosphere and ectomycorrhizal fungal communities were slightly to strongly undersampled, as the richness estimators predicted 58–90 % and 85–97 % inventory exhaustiveness, respectively, despite the analysis of 18,000–40,000 reads. Another caution that should be borne in mind in the analysis of fungal communities sampled with 454 pyrosequencing is that the number of 454 reads does not always reflect the abundance of templates of particular MOTUs in the samples examined (Amend et al. 2010a) but can be artificial due to PCR bias caused by mismatches and taxonomic affinities of primers (Bellemain et al. 2010).

1.6   DNA Barcoding of Fungi

DNA barcoding refers to a mechanism for rapid and accurate species identification based on a short standardized DNA sequence (usually 500–800 bp long) that corresponds to the same locus for all species included in particular taxonomic groups (Valentini et al. 2008). DNA barcoding has remarkable potential as a tool to facilitate identifying fungal species, as fungi are mostly inconspicuous or microscopic and have morphologically very similar species (Seifert 2009). The applicability of DNA barcoding has been confirmed for the identification of lichenized fungi (Kelly et al. 2011). The importance of DNA barcoding is becoming clear as the new high-throughput parallel pyrosequencing of fungal DNA from environmental samples becomes popular and as reliable taxonomic affiliations of numerous MOTUs are consequently necessitated (Tables 1.1 and 1.2). The usefulness of DNA barcoding of fungi largely depends on (1) the choice of genetic marker that is amplified with robust primers for fungal groups of particular interest or with universal primers for all fungal groups and (2) the development of high-quality sequence databases that contain fully identified sequences (FIS).

Several DNA regions have been used as genetic markers of fungi, such as ITS (including 5.8S) region, SSU, and LSU (including two variable subregions called D1 and D2) of rDNA (Fig. 1.2), β-tubulin, elongation factor 1a (EF1a), actin, and RNA polymerase II second largest subunit (RPB2). The difficulties of finding the appropriate marker(s) have not been completely resolved, but the ITS region of nrDNA is the preferred universal DNA barcoding marker of fungi commonly used for the identification of both single taxa and mixed environmental templates. The ITS region has such advantages as (1) it is easily amplifiable with universal primers (Bellemain et al. 2010; Eberhardt 2010), (2) it has resolution at various scales (ITS1, rapidly evolving; 5.8S, very conserved; ITS2 moderately rapidly to rapidly evolving) (Nilsson et al. 2008), and (3) a wealth of sequence data is available in databases (Begerow et al. 2010, see below). On the other hand, the ITS region has disadvantages of relatively short length of amplicons (500 bp on average; Seifert 2009) and potential PCR biases (Bellemain et al. 2010). The LSU has been the standard marker for identification of yeasts (Seifert 2009). Cytochrome c oxidase subunit I (COI) of mitochondrial DNA, a universal DNA barcode for animals and algae (Chase and

Fay 2009), was shown to be appropriate only for a few fungal groups, such as *Penicillium* (Seifert et al. 2007), and is believed to be unsuitable as a universal marker because multiple copies and introns have been reported (Seifert 2009; Begerow et al. 2010; Dentinger et al. 2011).

Previous studies are inconsistent with the usefulness of ITS as a MOTU identifier for ecological and biogeographical purposes. Taylor et al. (2006) claimed that molecular phylogenetic recognition of fungal species using an array of genetic markers including ITS is generally useful in distinguishing multiple endemic species within a single cosmopolitan fungal species complex defined by their morphology. Contrary to this, Gazis et al. (2011) performed multilocus phylogenetic analyses (*gpd*, ITS, and *tef1*) for fungal isolates of three endophytic fungal species complex and demonstrated that ITS alone underestimated the number of MOTUs predicted by the other nuclear loci and had low resolution to detect host specificity and geographical distribution.

The sequence length of amplicons is crucial for the reliability of DNA barcoding. Min and Hickey (2007) examined the effect of reducing the length of five fungal mitochondrial genes on the utility of the data for DNA barcoding. The reduction in sequence length diminished the accuracy of resulting phylogenetic trees but still yielded accurate species identification. It is unclear whether the same holds true for the sequence length of ITS amplicons.

Several databases are available online for fungal DNA barcoding (summarized in Seifert 2009). The INSD [GenBank, European Molecular Biology Laboratory (EMBL), and DNA Database of Japan (DDBJ)] is the major open repository for sequence data commonly used for fungi. A total of more than 2.4 M fungal sequences are deposited in the INSD, and ITS, SSU, and LSU of rDNA dominate in terms of the number of sequences deposited (Begerow et al. 2010). The yearly sequence submissions are increasing rapidly, especially since 2007. The INSD contains FIS of the ITS region for about 13,350 species (Nilsson et al. 2009), which corresponds to 13.6 % of the number of described species (0.1 M) and to 0.9 % of the hypothesized number (1.5 M) of fungal species. Specific databases or barcoding projects devoted to species of, for example, economically important fungi, such as *Penicillium*, *Fusarium*, and *Trichoderma* (Seifert 2009), European ectomycorrhizal fungi (UNITE, Abarenkov et al. 2010), and indoor and quarantine fungi (Groenewald 2009).

The low percentage of FIS of the ITS region indicates that insufficiently identified sequences (IIS) whose species affiliation remains unknown account for the major part of the database (Ryberg et al. 2008). Nilsson et al. (2008) noted that the deposition rate of IIS in the INSD now parallels (or exceeds) that of FIS. The majority of these IIS originate from environmental samples, and it is obvious that the increase of such IIS in the database decreases the chance of satisfactory hits via the Basic Local Alignment Search Tool (BLAST) to a sequence of known taxonomic identity (Hibbett et al. 2011). Consequently, there is a negative feedback in which the flood of IIS from NGS technologies leads to unsatisfactory BLAST hits to numerous IIS in the database. Development of modeling and bioinformatic approaches to annotate 454 sequence sequences with reference to "high-quality" databases will be urgently needed as an alternative of BLAST searching (e.g. Ovaskainen et al. 2010).

Unfortunately, the deposition rate of specimen-based FIS in the INSD is rather low. Only 20–40 % of the new fungal species described each year have data of DNA sequences of any locus in the database, whereas the remaining 60–80 % (600–1,000 new species each year) have no reference sequence data (Hibbett et al. 2011). In an analysis of specimens deposited at the fungal herbarium of the Royal Botanic Gardens of Kew (UK), Brock et al. (2008) reported that 69 % (192 of 279 sequences from the herbarium specimens) were not yet represented in the INSD with ≥97% sequence similarity, 19 % matched to FIS, and 12 % matched to IIS. The result of Brock et al. (2008) suggested that fully identified specimens in fungal herbariums could potentially contribute to the database for DNA barcoding, but many of them remained to be sequenced, partly due to poor conditions of specimen preservation. Nagy et al. (2011) pointed out the possibility that the unusually high number of unidentifiable MOTUs in environmental samples can be, at least in some fungal groups, ascribed to a lag in sequencing of type strain and specimen, rather than to a high number of undescribed species. They claimed that analysis of DNA sequences of reference collections would make a large contribution to the identification of fungal environmental sequences. However, the result of Nagy et al. (2011) does not exclude a possibility that many fungal species are yet to be described for other fungal groups than their focal genus *Mortierella* (Mucoromycotina). It is obviously needed to continue mycological observation, isolation, and description of novel fungal specimen and isolates, especially those of unidentified MOTUs detected with environmental DNA analyses. Further improvements of traditional mycological techniques, such as isolation and incubation procedures, are inevitable for these purposes (Fig. 1.1).

1.7   Tropical Regions As a Hotspot of Fungal Diversity

Tropical regions are known for their wealth of biodiversity under optimal temperature and moisture conditions for biological activities, and their long history of evolution (Corlett and Primack 2011). Recent studies have shown that the species richness of fungi is greater in warmer and wetter climates along climatic gradients for endophytic (Arnold and Lutzoni 2007), saprobic (Osono 2011b), and indoor fungi (Amend et al. 2010b), consistent with the general latitudinal gradient of biodiversity (Hillebrand 2004) [but see Tedersoo and Nara (2010) for the reverse pattern of latitudinal gradient of ectomycorrhizal fungal diversity]. Hawksworth (2001) stressed the need for further studies on fungal inventories in tropical regions, as they are a rich source of novel species. The hypothesized hyper-diversity and the mass of fungal species yet to be discovered and described merit the application of metagenomics and DNA barcoding techniques for the rapid and effective evaluation of tropical fungal diversity. Projects to integrate DNA barcoding into inventory programs for tropical biodiversity are ongoing for plants (Gonzalez et al. 2011) and insects (Janzen et al. 2009; Janzen and Hallwachs 2011). However, tropical mycology addresses the description of the α diversity of major taxonomic and functional groups (e.g. Watling et al. 2002a, b), and pyrosequencing of tropical fungal DNA in environmental samples has just started (Table 1.1). To the knowledge of the author, no publication is available regarding an inventory and DNA barcoding project of tropical fungi.

The author's group currently conducts research projects aimed at revealing fungal biodiversity and functioning in Asian regions. A core study site is established in the mountainous area in the northern part of Okinawa Island, in southern Japan (Osono 2011a). The subtropical forest in that area, called Yanbaru in Ryukyuan languages, is one of the biodiversity hot spots in Japan (Biodiversity Center of Japan 2010). Yanbaru is especially known for its richness in endemic birds, frogs, and land reptiles. Biodiversity assessments are urgently needed for microscopic organisms in this species-rich region, especially for fungi that play crucial roles in ecosystem functioning and services.

The biodiversity of litter- and wood-decomposing, mycorrhizal, endophytic, and soil fungi are investigated in the Yanbaru project (Osono et al. 2008; Fukasawa et al. 2012). Over the last decade, fieldwork has been carried out to collect fruiting bodies of fungi and live and dead tissues of plants and animals for fungal biodiversity assessments. The samples are taken back to the laboratory and used for the isolation of fungi and molecular phylogenetic analyses. Pyrosequencing of environmental samples such as live and dead leaves and soils have been performed for the last few years to test if fungal metagenomics is applicable to the description of putative hyper-diversity of tropical fungi. We are now analyzing more than 1.1 M fungal DNA sequences that a Roche 454 has generated. One project concerns endophytic fungi on healthy-looking live leaves of *Castanopsis sieboldii* and *Schima wallichii*, the dominant tree species in the study area. Preliminary analyses indicated that more than 190,000 reads of fungal ITS1 regions from the leaves were grouped into 350 MOTUs (with a threshold of 95 % similarity). Similarly, approximately 420,000 reads of fungal ITS regions retrieved from dead leaves, soils, and castings of earthworms were preliminarily grouped into 2,650 MOTUs, some of which showed phylogenetic affinity to mycorrhizal and litter-decomposing fungal species observed in the study area. These results imply a hyper-diverse nature of the subtropical fungal assemblage, compared to the MOTU richness previously reported in temperate forests (Table 1.1). Further studies are needed to examine the applicability of DNA barcoding of fungi and to reveal taxonomic aspects of fungal diversity in Yanbaru and other tropical regions in Asia.

1.8 Linking Fungal Diversity to Functioning

Pyrosequencing of the ITS regions of fungal DNA from environmental samples provides taxonomic profiles of the fungal diversity, but functional aspects of the fungal diversity cannot be thus explored. Putative functions of MOTUs may be inferred if their taxonomic positions are reasonably suggestive of specific functional groups, such as arbuscular mycorrhizal fungi in the Glomeromycota. However, there is no means of assuring whether a particular MOTU$_{ITS}$ takes part in metabolism or biological interaction in the environment, or is active or dormant. Because of the inherent difficulties involved in studying the function of minute mycelia of multiple fungal species in the environment, previous studies have estimated the functional aspects either at the level of individual fungal species indirectly with manipulative experiments using fungal isolates or at the level of entire fungal assemblages without taking the activity of individual fungal species into consideration (e.g. Osono 2007). Analysis of fungal RNA (complementary DNA or cDNA) will be fruitful in this respect as it can throw light upon the active component of fungal communities and their expression patterns in environmental samples. Adopting functional gene-encoding enzymes for specific metabolic pathways, such as heme peroxidases that degrade lignin (Morgenstern et al. 2008; Kellner et al. 2009) or phosphatases that hydrolyze phosphoric acid monoesters into phosphate ions (Sinsabaugh et al. 1991), as genetic markers will be helpful in evaluating the functional aspects of fungal diversity. Along with the molecular approaches to fungal functioning, future studies are clearly needed to verify the functioning of individual fungal species (or MOTUs) using fungal isolates.

Acknowledgments   I thank Dr. D. Hirose and Dr. E. Nakajima for critical reading of this manuscript; Mr. S. Matsuoka, Ms. C. Sakaguchi, Mr. K. Ito, Dr. S. Yazawa, Mr. O. Nishimura, Dr. H. Toju, and Dr. A. Tanabe for collaborations with pyrosequencing and bioinformatics. This work was supported by Global COE Program A06 of Kyoto University, the Japanese Ministry of Education, Culture and Sports (No. 23770083), a JGC-S Scholarship Foundation for Young Researchers, the New Technology Development Foundation, and Nippon Life Insurance Foundation.

References

Abarenkov K, Nilsson RH, Larsson KH, Alexander IJ, Eberhardt U, Erland S, Høiland K, Kjøller R, Larsson E, Pennanen T, Sen R, Taylor AFS, Tedersoo L, Ursing BM, Vrålstad T, Liimatainen K, Peintner U, Kõljalg (2010) The UNITE database for molecular identification of fungi – recent updates and future perspectives. New Phytol 186:281–285
Aly AH, Debbab A, Kjer J, Proksch P (2010) Fungal endophytes from higher plants: a prolific source of phytochemicals and other bioactive natural products. Fun Div 41:1–16
Amend AS, Seifert KA, Bruns TD (2010a) Quantifying microbial communities with 454 pyrosequencing: does read abundance count? Mol Ecol 19:5555–5565
Amend AS, Seifert KA, Samson R, Bruns TD (2010b) Indoor fungal composition is geographically patterned and more diverse in temperate zones than in the tropics. Proc Natl Acad Sci 107: 13748–13753
Arfi Y, Buée M, Marchand C, Levasseur A, Record E (2012) Multiple markers pyrosequencing reveals highly diverse and host-specific fungal communities on the mangrove trees *Avicennia marina* and *Rhizophora stylosa*. FEMS Microbiol Ecol 79:433–444
Arnold AE, Lutzoni F (2007) Diversity and host range of foliar fungal endophytes: are tropical leaves biodiversity hotspots? Ecology 88:541–549
Begerow D, Nilsson H, Unterseher M, Maier W (2010) Current state and perspectives of fungal DNA barcoding and rapid identification procedures. Appl Microbiol Biotechnol 87:99–108

Bellemain E, Carlsen T, Brochmann C, Coissac E, Taberlet P, Kauserud H (2010) ITS as an environmental DNA barcode for fungi: an *in silico* approach reveals potential PCR biases. BMC Microbiol 10:189

Biodiversity Center of Japan (2010) Biodiversity of Japan, a harmonious coexistence between nature and humankind. Heibonsha, Tokyo

Blackwell M (2011) The fungi: 1,2,3 … 5.1 million species? Am J Bot 98:426–438

Boddy L, Frankland J, van West P (2008) Ecology of saprotrophic basidiomycetes. Academic, London

Brock PM, Doring H, Bidartondo MI (2008) How to know unknown fungi: the role of a herbarium. New Phytol 181:719–724

Buée M, Reich M, Murat C, Morin E, Nilsson RH, Uroz S, Martin F (2009) 454 pyrosequencing analyses of forest soils reveal an unexpectedly high fungal diversity. New Phytol 184:449–456

Cavalier-Smith T (1987) The origin of fungi and pseudofungi. In: Rayner ADM, Chapman DJ (eds) Evolutionary biology of the fungi. Cambridge University Press, Cambridge

Chase MW, Fay MF (2009) Barcoding of plants and fungi. Science 325:682–683

Corlett RT, Primack RB (2011) Tropical rain forests: an ecological and biogeographical comparison. Wiley-Blackwell, Chinchester

Dentinger BTM, Didukh MY, Moncalvo JM (2011) Comparing COI and ITS as DNA barcode markers for mushrooms and allies (Agaricomycotina). PLoS One 6:e25081

Dumbrell AJ, Ashton PD, Aziz N, Feng G, Nelson M, Dytham C, Fitter AH, Helgason T (2011) Distinct seasonal assemblages of arbuscular mycorrhizal fungi revealed by massively parallel pyrosequencing. New Phytol 190:794–804

Eberhardt U (2010) A constructive step towards selecting a DNA barcode for fungi. New Phytol 187:265–268

Fukasawa Y, Osono T, Takeda H (2012) Fungal decomposition of woody debris of *Castanopsis sieboldii* in a subtropical old-growth forest. Ecol Res 27:211–218

Gams W (1992) The analysis of communities of saprophytic microfungi with special reference to soil fungi. In: Winterhoff W (ed) Fungi in vegetation science. Kluwer, Dordrecht, pp 182–223

Gazis R, Rehner S, Chaverri P (2011) Species delimitation in fungal endophyte diversity studies and its implications in ecological and biogeographic inferences. Mol Ecol 20:3001–3013

Ghannoum MA, Jurevic RJ, Mukherjee PK, Cui F, Sikaroodi M, Naqvi A, Gillevet PM (2010) Characterization of the oral fungal microbiome (mycobiome) in healthy individuals. PLoS Pathog 6:e1000713

Glenn TC (2011) Field guide to next-generation DNA sequencers. Mol Ecol Res 11:759–769

Goel R, Kotay SM, Butler CS, Torres CI, Mahendra S (2011) Molecular biological methods in environmental engineering. Water Environ Res 29:927–955

Gonzalez MA, Baraloto C, Engel J, Mori SA, Pétronelli P, Riéra B, Roger A, Thébaud C, Chave J (2011) Identification of Amazonian trees with DNA barcodes. PLoS One 4:e7483

Gottel NR, Castro HF, Kerley M, Yang Z, Pelletier DA, Podar M, Karpinets T, Uberbacher E, Tuskan GA, Vilgalys R, Doktycz MJ, Schadt CW (2011) Distinct microbial communities within the endosphere and rhizosphere of *Populus deltoides* roots across contrasting soil types. Appl Environ Microbiol 77:5934–5944

Groenewald JZ (2009) Update on fungal DNA barcoding campaigns. Persoonia 23:179

Hawksworth DL (1991) The fungal dimension of biodiversity: magnitude, significance, and conservation. Mycol Res 95:641–655

Hawksworth DL (2001) The magnitude of fungal diversity: the 1.5 million species estimate revisited. Mycol Res 105:1422–1432

Hibbett DS, Binder M, Bischoff JF, Blackwell M, Cannon PF, Eriksson OE, Huhndorf S, James T, Kirk PM, Lücking R, Lumbsch HT, Lutzoni F, Matheny PB, McLaughlin DJ, Powell MJ, Redhead S, Schoch CL, Spatafora JW, Stalpers JA, Vilgalys R, Aime MC, Aptroot A, Bauer R, Begerow D, Benny GL, Castlebury LA, Crous PW, Dai YC, Gams W, Geiser DM, Griffith GW, Gueidan C, Hawksworth DL, Hestmark G, Hosaka K, Humber RA, Hyde KD, Ironside JE,

Kõljalg U, Kurtzman CP, Larsson KH, Lichtwardt R, Longcore J, Miadlikowska J, Miller A, Moncalvo JM, Mozley-Standridge S, Oberwinkler F, Parmasto E, Reeb V, Rogers JD, Roux C, Ryvarden L, Sampaio JP, Schüßler A, Sugiyama J, Thorn RG, Tibell L, Untereiner WA, Walker C, Wang Z, Weir A, Weiss M, White MM, Winka K, Yao YJ, Zhang N (2007) A higher-level phylogenetic classification of the Fungi. Mycol Res 111:509–547

Hibbett DS, Ohman A, Kirk PM (2009) Fungal ecology catches fire. New Phytol 184:279–282

Hibbett DS, Ohman A, Glotzer D, Nuhn M, Kirk P, Nilsson RH (2011) Progress in molecular and morphological taxon discovery in Fungi and options for formal classification of environmental sequences. Fun Biol Rev 25:38–47

Hillebrand H (2004) On the generality of the latitudinal diversity gradient. Am Nat 163:192–211

Hughes KW, Petersen RH, Lickey EB (2009) Using heterozygosity to estimate a percentage DNA sequence similarity for environmental species' delimitation across basidiomycete fungi. New Phytol 182:795–798

Hui N, Jumpponen A, Niskanen T, Liimatainen K, Jones KL, Koivula T, Romantschuk M, Strömmer R (2011) EcM fungal community structure, but not diversity, altered in a Pb-contaminated shooting range in a boreal coniferous forest site in Southern Finland. FEMS Microbiol Ecol 76:121–132

James TY, Kauff F, Schoch CL, Matheny PB, Hofstetter V, Cox CJ, Celio G, Gueidan C, Fraker E, Miadlikowska J, Lumbsch HT, Rauhut A, Reeb V, Arnold AE, Amtoft A, Stajich JE, Hosaka K, Sung GH, Johnson D, O'Rourke B, Crockett M, Binder M, Curtis JM, Slot JC, Wang Z, Wilson AW, Schüßler A, Longcore JE, O'Donnell K, Mozley-Standridge S, Porter D, Letcher PM, Powell MJ, Taylor JW, White MM, Griffith GW, Davies DR, Humber RA, Morton JB, Sugiyama J, Rossman AY, Rogers JD, Pfister DH, Hewitt D, Hansen K, Hambleton S, Shoemaker RA, Kohlmeyer J, Volkmann-Kohlmeyer B, Spotts RA, Serdani M, Crous PW, Hughes KW, Matsuura K, Langer E, Langer G, Untereiner WA, Lücking R, Büdel B, Geiser DM, Aptroot A, Diederich P, Schmitt I, Schultz M, Yahr R, Hibbett DS, Lutzoni F, McLaughlin DJ, Spatafora JW, Vilgalys R (2006) Reconstructing the early evolution of Fungi using a six-gene phylogeny. Nature 443:818–822

Janzen DH, Hallwachs W (2011) Joining inventory by parataxonomists with DNA barcoding of a large complex tropical conserved wildland in Northweseetern Costa Rica. PLoS One 6:e18123

Janzen DH, Hallwachs W, Blandin P, Burns JM, Cadiou JM, Chacon I, Dapkey T, Deans AR, Epstein ME, Espinoza B, Franclemont JG, Haber WA, Hajibabaei M, Hall JW, Hebert PDN, Gauld ID, Harvey DJ, Hausmann A, Kitching IJ, Lafontaine D, Landry JF, Lemaire C, Miller JY, Miller JS, Miller L, Miller SE, Montero J, Munroe E, RabGreen S, Ratnasingham S, Rawlins JE, Robbins RK, Rodriguez JJ, Rougerie R, Sharkey MJ, Smith MA, Solis MA, Sullivan JB, Thiaucourt P, Wahl DB, Weller SJ, Whitfield JB, Willmott KR, Wood DM, Woodley NE, Wilson JJ (2009) Integration of DNA barcoding into an ongoing inventory of complex tropical biodiversity. Mol Ecol Resour 9:1–26

Jennings DH, Lysek G (1999) Fungal biology, understanding the fungal life style. BIOS, Oxford

Jumpponen A, Jones KL (2009) Massively parallel 454 sequencing indicates hyperdiverse fungal communities in temperate *Quercus macrocarpa* phyllosphere. New Phytol 184:438–448

Jumpponen A, Jones KL (2010) Seasonally dynamic fungal communities in the *Quercus macrocarpa* phyllosphere differ between urban and nonurban environments. New Phytol 186:496–513

Jumpponen A, Jones KL, Mattox D, Yaege C (2010a) Massively parallel 454-sequencing of fungal communities in *Quercus* spp. ectomycorrhizas indicates seasonal dynamics in urban and rural sites. Mol Ecol 19:41–53

Jumpponen A, Jones KL, Blair J (2010b) Vertical distribution of fungal communities in tallgrass prairie soil. Mycologia 102:1027–1041

Kellner H, Luis P, Schlitt B, Buscot F (2009) Temporal changes in diversity and expression patterns of fungal laccase genes within the organic horizon of a brown forest soil. Soil Biol Biochem 41:1380–1389

Kelly LJ, Hollingsworth PM, Coppins BJ, Ellis CJ, Harrold P, Tosh J, Yahr R (2011) DNA barcoding of lichenized fungi demonstrates high identification success in a floristic context. New Phytol 191:288–300

Kirk P, Cannon P, Stalpers J (2008) Dictionary of the fungi, 10th edn. CABI, Wallingford

Lekberg Y, Schnoor T, Kjøller R, Gibbons SM, Hansen LH, Al-Soud WA, Sørensen SJ, Rosendahl S (2012) 454-sequencing reveals stochastic local reassembly and high disturbance tolerance within arbuscular mycorrhizal fungal communities. J Ecol 100:151–160

Lentendu G, Zinger L, Manel S, Coissac E, Choler P, Geremia RA, Melodelima C (2011) Assessment of soil fungal diversity in different alpine tundra habitats by means of pyrosequencing. Fun Div 49:113–123

Liggenstoffer AS, Youssef NH, Couger MB, Elshahed MS (2010) Phylogenetic diversity and community structure of anaerobic gut fungi (phyum Neocallimastigomycota) in ruminant and non-ruminant herbivores. ISME J 4:1225–1235

Lim YW, Kim BK, Kim C, Jung HS, Kim BS, Lee JH (2010) Assessment of soil fungal communities using pyrosequencing. J Microbiol 48:284–289

Lucero ME, Unc A, Cooke P, Dowd S, Sun S (2011) Endophyte microbiome diversity in micro-propagated *Striplex canescens* and *Atriplex torreyi* var *griffithsii*. PLoS One 6:e17693

Lumini E, Orgiazzi A, Borriello R, Bonfante P, Bianciotto V (2010) Disclosing arbuscular mycorrhizal fungal biodiversity in soil through a land-use gradient using a pyrosequencing approach. Environ Microbiol 12:2165–2179

Lutzoni F, Miadlikowska J (2009) Lichens. Curr Biol 19:502–503

Malausa T, Gilles A, Meglécz E, Blanquart H, Duthoy S, Costedoat C, Dubut V, Pech N, Castagnone-sereno P, Délye C, Feau N, Frey P, Gauthier P, Guillemaud T, Hazard L, Le Corre V, Lung-Escarmant B, Malé PJG, Ferreira S, Martin JF (2011) High-throughput microsatellite isolation through 454 GS-FLX Titanium pyrosequencing of enriched DNA libraries. Mol Ecol Res 11:638–644

Mello A, Napoli C, Murat C, Morin E, Marceddu G, Bonfante P (2011) ITS-1 versus ITS-2 pyrosequencing: a comparison of fungal populations in truffle grounds. Mycologia 103:1184–1193

Metzker ML (2010) Sequencing technologies – the next generation. Nat Rev Genet 11:31–46

Min XJ, Hickey DA (2007) Assessing the effect of varing sequence length on DNA barcoding of fungi. Mol Ecol Notes 7:365–373

Monchy S, Sanciu G, Jobard M, Rasconi S, Gerphagnon M, Chabé M, Cian A, Meloni D, Niquil N, Christaki U, Viscogliosi E, Sime-Ngando T (2011) Exploring and quantifying fungal diversity in freshwater lake ecosystems using rDNA cloning/sequencing and SSU tag pyrosequencing. Environ Microbiol 113:1433–1453

Moora M, Berger S, Davison J, Öpik M, Bommarco R, Bruelheide H, Kühn I, Kunin WE, Metsis M, Rortais A, Vanatoa A, Vanatoa E, Stout JC, Truusa M, Westphal C, Zobel M, Walther GR (2011) Alien plants associate with widespread generalist arbuscular mycorrhizal fungal taxa: evidence from a continental-scale study using massively parallel 454 sequencing. J Biogeogr 38:1305–1317

Mora C, Tittensor DP, Adl S, Simpson AGB, Worm B (2011) How many species are there on Earth and in the ocean? PLoS Biol 9:e1001127

Morgenstern I, Klopman S, Hibbett DS (2008) Molecular evolution and diversity of lignin degrading heme peroxidases in the Agaricomycetes. J Mol Evol 66:243–257

Nagy LG, Petkovits T, Kovács GM, Voigt K, Vágvölgyi C, Papp T (2011) Where is the unseen fungal diversity hidden? A study of *Mortierella* reveals a large contribution of reference collections to the identification of fungal environmental sequences. New Phytol 191:789–794

Nilsson RH, Kristiansson E, Ryberg M, Hallenberg N, Larsson KH (2008) Intraspecific ITS variability in the kingdom Fungi as expressed in the international sequence databases and its implications for molecular species identification. Evol Bioinform online 4:193–201

Nilsson RH, Ryberg M, Abarenkov K, Sjökvist E, Kristiansson E (2009) The ITS region as a target for characterization of fungal communities using emerging sequencing technologies. FEMS Microbiol Lett 296:97–101

Nilsson RH, Tedersoo L, Lindahl BD, Kjøller R, Carlsen T, Quince C, Abarenkov K, Pennanen T, Stenlid J, Bruns T, Larsson KH, Kõljalg U, Kauserud H (2011) Towards standardization of the description and publication of next-generation sequencing datasets of fungal communities. New Phytol 191:314–318

Nowrousian M (2010) Next-generation sequencing techniques for eukaryotic microorganisms: sequencing-based solution to biological problems. Eukaryot Cell 9:1300–1310

O'Brien HE, Parrent JL, Jackson JA, Moncalvo JM, Vilgalys R (2005) Fungal community analysis by large-scale sequencing of environmental samples. Appl Environ Microbiol 71:5544–5550

Okubo A, Sugiyama S (2009) Comparison of molecular fingerprinting methods for analysis of soil microbial community structure. Ecol Res 24:1399–1405

Öpik M, Metsis M, Daniell TJ, Zobel M, Moora M (2009) Large-scale parallel 454 sequencing reveals host ecological group specificity of arbuscular mycorrhizal fungi in a boreonemoral forest. New Phytol 184:424–437

Osono T (2007) Ecology of ligninolytic fungi associated with leaf litter decomposition. Ecol Res 22:955–974

Osono T (2011a) Yanbaru fungal biodiversity project. DIWPA Newslett 25:8–9

Osono T (2011b) Diversity and functioning of fungi associated with leaf litter decomposition in Asian forests of different climatic regions. Fun Ecol 4:375–385

Osono T, Ishii Y, Hirose D (2008) Fungal colonization and decomposition of *Castanopsis sieboldii* leaf litter in a subtropical forest. Ecol Res 23:909–917

Ovaskainen O, Nokso-Koivisto J, Hottola J, Rajala T, Pennanen T, Ali-Kovero H, Miettinen O, Oinonen P, Auvinen P, Paulin L, Larsson KH, Mäkipää R (2010) Identifying wood-inhabiting fungi with 454 sequencing - what is the probability that BLAST gives the correct species? Fun Ecol 3:274–283

Rousk J, Baath E, Brookes PC, Lauber CL, Lozupone C, Caporaso JG, Knight R, Fierer N (2010) Soil bacterial and fungal communities across a pH gradient in an arable soil. ISME J 4:1340–1351

Ryberg M, Kristiansson E, Sjökvist E, Nilsson RH (2008) An outlook on the fungal internal transcribed spacer sequences in GenBank and the introduction of a web-based tool for the exploration of fungal diversity. New Phytol 181:471–477

Schmit JP, Mueller GM (2007) An estimate of the lower limit of global fungal diversity. Biodivers Conserv 16:99–111

Seifert KA (2009) Progress towards DNA barcoding of fungi. Mol Ecol Res 9:83–89

Seifert KA, Samson RA, Dewaard JR, Houbraken J, Levesque CA, Moncalvo JM, Louis-Seize G, Hebert PDN (2007) Prospects for fungus identification using COI DNA barcodes, with *Penicillium* as a test case. Proc Natl Acad Sci USA 104:3901–3906

Sinsabaugh RL (2010) Phenol oxidase, peroxidase and organic matter dynamics in soil. Soil Biol Biochem 42:391–404

Sinsabaugh RL, Antibus RK, Linkins AE (1991) An enzymic approach to the analysis of microbial activity during plant litter decomposition. Agric Ecosyst Environ 34:43–54

Stajich JE, Berbee ML, Blackwell M, Hibbett DS, James TY, Spatafora JW, Taylor JW (2009) Fungi. Curr Biol 19:840–845

Su C, Lei L, Duan Y, Zhang KQ, Yang J (2012) Culture-independent methods for studying environmental microorganisms: methods, application, and perspective. Appl Mirobiol Biotechnol 93:993–1003

Taylor JW, Berbee ML (2006) Dating divergences in the Fungal Tree of Life: review and new analyses. Mycologia 98:838–849

Taylor JW, Turner E, Townsend JP, Dettman JR, Jacobson D (2006) Eukaryotic microbes, species recognition and the geographic limits of species: examples from the kingdom Fungi. Phil Trans R Soc B 361:1947–1963

Tedersoo L, Nara K (2010) General latitudinal gradient of biodiversity is reversed in ectomycorrhizal fungi. New Phytol 185:351–354

Tedersoo L, Nilsson RH, Abarenkov K, Jairus T, Sadam A, Saar I, Bahram M, Bechem E, Chuyong G, Kõljalg U (2010) 454 pyrosequencing and Sanger sequencing of tropical mycorrhizal fungi provide similar results but reveal substantial metholodogical biases. New Phytol 188:291–301

Unterseher M, Jumpponen A, Öpik M, Tedersoo L, Moora M, Dormann CF, Schnittler M (2011) Species abundance distribution and richness estimations in fungal metagenomics - lessons learned from community ecology. Mol Ecol 20:275–285

Valentini A, Pompanon F, Taberlet P (2008) DNA bacoding for ecologists. Trends Ecol Evol 24:110–117

Voigt K, Kirk PM (2011) Recent developments in the taxonomic affiliation and phylogenetic positioning of fungi: impact in applied microbiology and environmental biotechnology. Appl Microbiol Biotechnol 90:41–57

Wallander H, Johansson U, Sterkenburg E, Durling MB, Lindahl BD (2010) Production of ecto-mycorrhizal mycelium peaks during canopy closure in Norway spruce forests. New Phytol 187:1124–1134

Watling R, Frankland JC, Ainsworth AM, Issac S, Robinson CH (2002a) Tropical Mycology. Vol. 1. Macromycetes. CABI Publishing, Oxon

Watling R, Frankland JC, Ainsworth AM, Issac S, Robinson CH (2002b) Tropical Mycology. Vol. 1. Micromycetes. CABI Publishing, Oxon

Zamocky M, Obinger C (2010) Molecular phylogeny of heme peroxide. In: Torres E, Ayala M (eds) Biocatalysis based on heme peroxidases. Springer, Berlin

Chapter 2
Insect–Plant Interactions in Plant-based Community/Ecosystem Genetics

Abstract Plant traits are fundamental for characterize population, community, and ecosystem properties in a terrestrial ecosystem. Recently, there is increasing evidence that genetically controlled plant traits play an important role in determining community structure and ecosystem processes (i.e., community/ecosystem genetics). On the other hand, we should recognize that herbivores modify plant traits and greatly influence their impacts in determining such community and ecosystem pro perties through direct and indirect interactions. Here, we review the accumulated knowledge of community/ecosystem genetics, and highlight how herbivores are important for modifying the community and ecosystem consequences of genetically controlled plant traits. Our review clearly illustrates that genetically controlled and plastically modified characteristics of plants are undoubtedly important determinants of community and ecosystem properties. In particular, most such plant traits that influence community and ecosystem properties are related to antiherbivore defense. Therefore, not only genetically controlled defensive traits of plants but also plastically modified defensive traits should be taken into consideration to understand the evolution of antiherbivore defensive traits of plants in community and ecosystem contexts.

Keywords Antiherbivore defense • Community structure • Ecosystem process • Herbivore-induced plant response

2.1 Introduction

Plants are the important key organisms that support food webs and material cycling in terrestrial ecosystems. A critical role of plant characteristics has been classically recognized in shaping community structures of associated arthropods (Feeny 1970) and in regulating ecosystem processes (Aerts 1997). Recently, researchers have tried to reveal the evolutionary process of such plant characteristics by focusing on their genetic background, which reciprocally interacts with community and

T. Sota et al., *Species Diversity and Community Structure: Novel Patterns and Processes in Plants, Insects, and Fungi*, SpringerBriefs in Biology, DOI 10.1007/978-4-431-54261-2_2, © The Author(s) 2014

ecosystem properties (i.e. community/ecosystem genetics, Whitham et al. 2006). On the other hand, there is accumulating evidence that herbivores can modify plant characteristics through their feeding behavior (Ohgushi 2005). In this context, herbivores have two important functions in determining the plant characteristics. One is function as a selective pressure for evolution of the plant characteristics (e.g. Agrawal 2006), and the other is a function as an inducer of changes in plant phenotypes (e.g. Karban and Baldwin 1997). Furthermore, plant traits also depend on soil nutrient availability (Bryant et al. 1983) which is potentially influenced by herbivores through the direct effects of excretion of their waste products (Lovett et al. 2002) and indirect effects of changes in the quality and quantity of plant litter (Chapman 2006). This recent recognition strongly indicates that we should incorporate a role of herbivores into the perspective of plant-based community/ecosystem genetics, to fully understand how plant characteristics link ecological processes in community and ecosystem, and how community or ecosystem properties in turn promote evolution of such plant characteristics.

In this chapter, we review the accumulated knowledge of plant-based community/ecosystem genetics, and highlight how herbivores, especially insect herbivores, are important for modifying the community and ecosystem consequences of plant traits. First, we summarize the importance of herbivore-induced plant responses as well as plant genotype in determining plant traits, and consequently the structure of insect communities. Second, we review how inter- and intra-specific variations in antiherbivore defenses of plants are important for determining ecosystem processes, such as litter decomposition and nutrient cycling, through effects on the feeding behavior of herbivores.

2.2 Community Consequences of Insect–Plant Interactions

2.2.1 Herbivore-Induced Plant Responses and Their Genetic Variation

Organisms display phenotypic plasticity, which is an ability of a single genotype to express multiple phenotypes in response to abiotic and/or biotic environmental conditions such as temperature, drought, predation pressure, and competition (Agrawal 2001; Dewitt and Schiner 2004; Miner et al. 2005). Phenotypic plasticity is important for adaptive solutions to cope with heterogeneous and unpredictable environments, and is ubiquitous across diverse organisms. One of the important kinds of phenotypic plasticity in a community context is trait change in response to species interactions. This is because such plasticity in response to species interactions inevitably emerges in community organization processes to a greater or less extent.

Terrestrial plants commonly develop a wide range of phenotypic responses to attack by herbivores (i.e., herbivore-induced plant responses), and the herbivore-induced plant responses as well as genetic variation in the host plants can play an important role in generating phenotypic variation (Ohgushi 2005; Utsumi et al. 2010; Utsumi 2011). The herbivore-induced plant responses can shape the arthropod community structure through changes in resource quality and quantity (Van Zandt and Agrawal 2004; Utsumi and Ohgushi 2009; Utsumi et al. 2009; Poelman et al. 2010; Ando et al. 2011a).

On the other hand, accumulating studies have shown that the structure of plant-associated arthropod communities differs among genetically distinct plants (Whitham et al. 2006). Community consequences of herbivore-induced plant responses and of host plant genetics for structuring plant-associated arthropod communities are likely to be qualitatively similar in terms of the profound impacts of plant characteristics on ecological communities. However, to date, these two aspects have been explored separately, and thus their relationships are poorly understood. Here, we aim to synthesize how induced plant responses and host plant genetics are involved in shaping ecological communities. Specifically, we focus on a key role of herbivore-induced phenotypic changes in terrestrial plants in creating variation of arthropod communities among genetically distinct host plants. This provides some insight into the mechanisms of the linkage between host plant genetics and associated communities.

Herbivory triggers plastic responses in morphology, reproduction, and tissue chemistry in plants by inducing resistance or susceptibility (Karban and Baldwin 1997; Ohgushi 2005). Such responses can be specific to a particular herbivore species, as indicated by the differential responses of plants to attack by different herbivores (McAuslane and Alborn 1998; Geervliet et al. 1997; Agrawal 2011). For example, induced responses in a wild radish, *Raphanus raphanistrum*, exhibit specificity for the type of herbivore, which shows differential induced resistance following attack by four lepidopteran herbivores (Agrawal 2000), and minimal induction following artificial clipping by scissors (Agrawal 1998; Agrawal et al. 1999). Responses to actual herbivory often differ qualitatively and/or quantitatively from responses to artificial defoliation (e.g., Hartley and Lawton 1987; Krause and Raffa 1992; Turlings et al. 1995; Stout et al. 1998). Several studies demonstrated that application of herbivore regurgitant to artificially damaged plant tissues can simulate more closely actual feeding, due to chemical elicitors present in the saliva (Lin et al. 1990; Alborn et al. 1997; Korth and Dixon 1997; McCloud and Baldwin 1997).

Plants also display tolerance: an ability to withstand herbivory with minimal losses with respect to growth and reproduction (Stowe et al. 2000; Utsumi and Ohgushi 2007). A compensatory response, which is regrowth and/or reproduction following herbivory to compensate for damaged tissue (Strauss and Agrawal 1999; Utsumi and Ohgushi 2007), is one of the most common and widespread plant tolerance strategies (Rosenthal and Kotanen 1994). The intensity of compensatory responses following natural herbivory often differs from that following artificial damage. Utsumi and Ohgushi (2007) showed that compensatory regrowth in lateral shoots of willows was more strongly enhanced by boring of a swift moth caterpillar than by artificial boring. Like plant resistance, plant compensatory responses may be affected by factors relevant to natural herbivory, such as specific feeding features or chemical cues in herbivore saliva (McNaughton 1983; Moon et al. 1994).

The above evidence suggests that plants evolve fine-tuned adaptive reactions to attack by a diverse array of insects. In fact, induced responses to insect herbivores have been shown to be genetically variable and heritable (Zangerl and Berenbaum 1990; van Dam and Vrieling 1994; English-Loeb et al. 1998), although genetic variation in induced responses to herbivory has been poorly explored. Juenger and Bergelson (2000) demonstrated a significant additive genetic variation in changes in flowering phenology and branch production of a scarlet gilia in response to herbivory. There was also a marginally significant additive genetic variance in fitness of herbivore-damaged plants, but not in fitness of control plants. Agrawal et al. (2002) reported additive genetic variation in inducibility of defensive traits (i.e., plasticity of glucosinolate concentration) in wild radish and in induced resistance to a specialist herbivore, *Pieris rapae*. These results suggest that herbivore-induced plant responses are subjected to natural selection imposed by herbivores. Interestingly, Agrawal et al. (2002) did not detect genetic variation in glucosinolate concentration under undamaged condition, while there was significant additive genetic variation under damaged conditions. They argued that plants might allocate little to constitutive defense due to high costs, resulting in a lack of genetic variation in defensive traits without herbivory. Juenger and Bergelson (2000) also reported similar results in compensatory responses to herbivory. Genetic variation in induced plant responses may be widespread in natural plant populations (Zangerl and Berenbaum 1990; van Dam and Vrieling 1994; Agrawal et al. 2002; Bingham and Agrawal 2010; Snoeren et al. 2010).

Overall, plant traits should be ubiquitously controlled by genotype-by-environment interactions ($G \times E$), in which the presence or absence of a particular herbivore can alter the mean and variance of a phenotype within and between plant populations through herbivore-induced plant responses.

2.2.2 Community-Level Consequences of Genetic Variation in Herbivore-Induced Responses

The induced plant response to herbivory is ubiquitous phenomenon in terrestrial ecosystems. Furthermore, genetic variation in such responses is likely to be common in natural plant populations. Thus, considering that herbivores affect plant phenotypes through $G \times E$ (genetics $\times$ environment) functions, the effects of plant genetics on ecological communities at higher trophic levels are more likely to be dependent on genetic variation in herbivore-induced plant responses.

We can consider two scenarios of how plant genotypic effects and induced plant response effects propagate to the community in (1) additive and (2) non-additive manners (Fig. 2.1). If there is no genetic variation in inducibility in important traits relevant to shaping communities, additive scenarios are predicted. If there is genetic variation in the inducibility in functional traits, non-additive scenarios are predicted.

First, we explain the additive scenarios with one empirical example. Using a hybrid system of three genetic classes of willows, Hochwender et al. (2005) conducted a hand-clipping treatment to mimic browsing by mammals. Both effects of

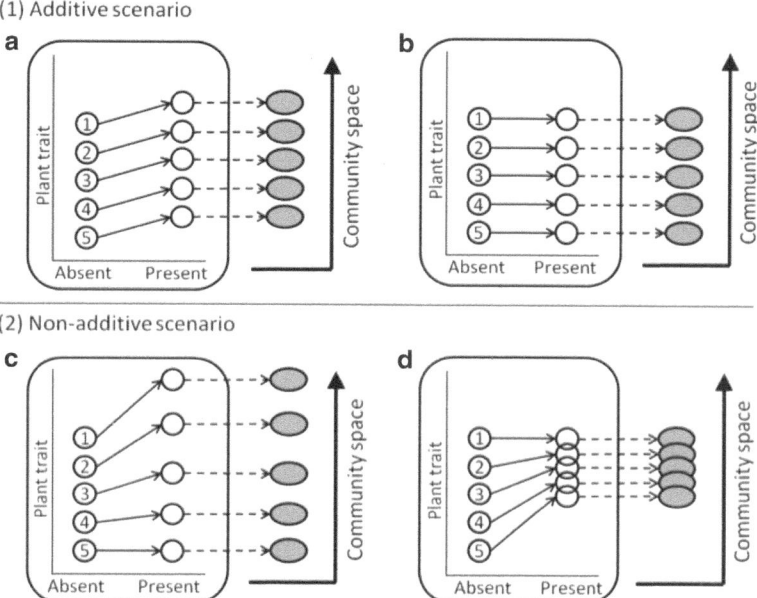

Fig. 2.1 Additive and non-additive scenarios for the interactions between effects of plant genetics and environmental effects of presence/absence of a particular herbivory on arthropod communities. (**1**) In additive scenarios, in a case of no genetic variation in inducibility in plant traits relevant to shaping communities, variation in arthropod communities among plant genotypes reflects the genetically determined variation in other traits among plant genotypes. In this case, two outcomes are predicted for community structure: (*a*) plant genotypes and herbivory additively influence plant traits and community structure, and (*b*) plant traits and resultant community structure are determined only by plant genotypic effects. (**2**) On the other hand, when there is significant genetic variation in inducibility, plant genetics and herbivory non-additively determine arthropod communities. In this case, two outcomes are also predicted. Variation in arthropod communities among plant genotypes can be (*c*) amplified or (*d*) mitigated by effects of herbivory, depending on a manner of genetic variation in inducibility (see text for details). Different numbers indicate distinct plant genotypes

genetic classes and clipping affected herbivore community structure additively. The clipping treatment resulted in longer shoot length, but the trait change did not differ among genetic classes. Non-significant genetic variation in changes in shoot length may explain the additive effect of genetic class and clipping effect (Fig. 2.1a). However, it should be noted that this study applied the artificial damage treatment. Artificial damage often triggers plant responses that differ from those caused by damage by natural herbivores, which are "real" agents of natural selection, and this difference might account for the lack of genetic variation in the willow responses. In addition, if plant traits important for shaping communities change to only a small extent in response to herbivory, genetic effects should be constant regardless of the presence or absence of herbivory (Fig. 2.1b). However, to our knowledge, no such patterns have been reported.

Next, we briefly introduce our ongoing field experiments to explain the non-additive scenario. We have investigated the plant-associated arthropod community on tall goldenrods, *Solidago altissima*, in a factorial design using multiple genotypes of tall goldenrod and aphid removal treatment. We genotyped tall goldenrods collected from the field, using amplified fragment length polymorphism (AFLP; Ando et al. unpublished data), and clonally propagated them to create replicates of the genotypes. On tall goldenrods, a specialist aphid plays an important role as a keystone herbivore that determines arthropod community structure through both ant- and plant-mediated indirect effects (Ando et al. 2011a, b). Aphid feeding can induce changes in plant traits: an increase in production of lateral shoots and new flush leaves and changes in plant quality such as nutrition traits (Ando and Ohgushi 2008), resulting in great impact on temporally separated arthropod communities. We found that tall goldenrods have genetic variation of resistance against the aphid (Utsumi et al. 2011), and that community composition of arthropod species significantly differed among plant genotypes in the presence and absence of the aphids. However, genotypes and aphid removal interactively influence community composition, and variation in community composition among plant genotypes was larger in the presence than in the absence of aphids. We also detected genotypic variation in aphid-induced plant responses, including lateral shoot production. There was also an among-genotype difference in volatile compounds emitted in response to leaf-clipping. Thus, the aphids may amplify variation in community composition due to genotypic differences in herbivore-induced plant responses (Fig. 2.1c). Furthermore, the genetic similarity in plant genotypes was closely related to the similarity in herbivore community composition on aphid-exposed plants. In contrast, no such relationships between plant genotypes and community structure were found on aphid-free plants. This suggests that genotypic variation in plant traits involved in induced responses to herbivory may be more important for variation in community structure among plant genotypes rather than other plant traits. In other words, there is a possibility that G×E amplified intraspecific trait variation among tall goldenrod individuals and the aphid-induced plant responses extended the plant genotypic effects to community-level through an aphid-induced indirect interaction web (Ohgushi et al. 2011).

In contrast, even when there is variation in plant inducibility among genotypes, trait values in some cases may converge to a similar level following herbivory. As a result, plant genotypic effects on community structure may be masked and not statistically detected (Fig. 2.1d). To clearly understand the relationship between the effects of plant phenotypic plasticity induced by herbivory and the genotypic variation among plants on the organization of arthropod communities, we need to accumulate evidence on how intraspecific genetic variation in herbivore-induced phenotypic plasticity can affect arthropod community structure among plants. However, to date no studies have explicitly addressed the question of how intraspecific genetic variation in plants and herbivore-induced responses interactively influence community properties. Only a few studies have shown much greater variation among plant genotypes in effects of induced plant responses on a single herbivore species than genotypic variation in other constitutive traits (Underwood et al. 2000).

Our common garden experiment with *Solidago* genotypes and aphid manipulation is also consistent with this pattern. Thus, a non-additive scenario may be more plausible in nature.

Because only manipulating plant genotypes in experiments cannot identify mechanisms by which plant genotypes influence arthropod communities, we need to conduct a multi-factorial experiment with the manipulation of the presence/absence of a particular herbivore species.

2.2.3 Insect–Plant Interactions Modifying Plant Genetic Effects

In the above section, we synthesize how plant genetics and the presence of herbivory interactively determine arthropod community structure. Furthermore, we discuss how other biological conditions in plant–insect interactions can influence plant genetic effects on shaping arthropod communities.

The spatial distribution and composition of plant genotypes can alter effects of a particular genotype on associated communities through insect behavior. The local genotypic diversity of plants affects herbivore movement, colonization, and emigration (Power 1988). Utsumi et al. (2011) showed that the synergistic effects of genotypes of tall goldenrods increased aphid population size in a diverse genotype patch more than expected from additive effects alone. As one possible explanation for this pattern, they suggested a source-sink relationship among genotypes of tall goldenrods: aphids move from genotypes with high reproductive success to genotypes with low reproductive success. Hence, plant genetic effects on the arthropod community can be greatly modified through interaction between the local plant genetic diversity and herbivore movement.

Genetic variation within both a plant species and an insect species often alters species interaction outcomes, leading to genotype-by-genotype interactions. Such $G \times G$ interactions can also be influenced by the presence or absence of the third species, thus resulting in genotype × genotype × environment interactions (i.e., $G \times G \times E$, Thompson 2005). For example, there was a significant interaction effect between aphid genotype and host plant genotype on the aphid abundance, while this interaction was also dependent on the presence or absence of rhizosphere bacteria (Tétard-Jones et al. 2007). Moreover, more recent studies have shown that different induction levels of plant responses to initial herbivory cause evolutionary shifts of traits of other herbivore species among populations (Utsumi et al. 2009; Utsumi et al. unpublished data). Such trait evolution of herbivores can also influence arthropod community structure through plant induced responses (Utsumi 2011; Utsumi et al. unpublished data). Thus, the $G \times G \times E$ framework may provide insights into a relationship between diffuse (co)evolution [i.e., selection or the response to selection imposed by one species on another, depending on the presence or absence of other species in a community (Strauss et al. 2005)] and shaping community structure. Future studies should focus not only on $G \times E$ but also on $G \times G \times E$.

2.2.4 Evolutionary Feedback to Induced Plant Defense

We have reviewed studies showing that herbivore-induced plant responses have considerable impacts on structuring arthropod communities and have a key function to link host plant genetics to associated communities. The reciprocal question of whether changes in arthropod community structure affect the evolution of plant traits remains unanswered.

In general, it is widely accepted that herbivores act as a selective force for plant defense and inducibility of such defense can be favored because of the reduced allocation of resources to defense when it is not needed (the allocation cost hypothesis) (Simms 1992). Moreover, the context of the ecological communities of herbivores may also favor inducibility, although this is not mutually exclusive with the allocation cost hypothesis. Most plants are attacked by diverse herbivores and have a high specificity of induced plant responses to herbivory. This specificity exhibits specific elicitation, in which changes in plant traits are dependent on the identity of the damaging agents, as we mentioned above, and also exhibits specific effects, i.e., different herbivore species display differential responses to a particular plant response (Utsumi et al. 2010; Agrawal 2011). Although there has been little exploration of how such specificity has evolved, such specificity of elicitation and effects may have evolved under selection imposed by communities consisting of a diverse range of herbivores. In other words, if genetic variation in specific induced plant responses commonly exists, it leads to shaping different community structures of herbivores among plant genotypes, which in turn, may impose selection for the induced plant responses. However, colonization of initial herbivore species and the response of herbivore communities could significantly vary year-to-year and site-to-site, depending on biotic conditions, as discussed above. Such variability in arthropod communities may lead to not only considerable ubiquity of induced plant responses but also the maintenance of their high genetic variation.

Therefore, we advocate a focus on natural genetic variation in induced plant responses and its community-wide consequences to study how variable arthropod communities result in diverse plant traits and how plant defense evolves. Synthesizing genetics and phenotypic plasticity in plants will provide profound insight toward an understanding ecological and evolutionary dynamics in plant-insect interactions.

2.3 Ecosystem Consequences of Insect–Plant Interactions

2.3.1 Insect Herbivores and Ecosystem Processes

Insect herbivores can influence ecosystem processes, such as decomposition and soil nutrient dynamics, through several mechanisms in terrestrial systems (Hunter 2001; Hartley and Jones 2004). They include (1) deposition of excrement, (2) inputs of cadavers, (3) changes in through-fall chemicals, (4) changes in quality and quantity of litter inputs, (5) changes in plant community structure, (6) changes in root exudation, and (7) changes in soil microclimates. Coupled with soil biological and abiological conditions, these mechanisms result in variable (accelerated, decelerated, or neutral) effects of insect herbivores on decomposition processes and soil nutrient dynamics (Stadler et al. 2004; Kay et al. 2008; Kagata and Ohgushi 2012).

The relationship between leaf palatability to insect herbivores and litter decomposability is one of the important factors determining the direction of effects of insect herbivores on ecosystem processes (Bardgett and Wardle 2010). Several studies have shown that plants with foliage more palatable to generalist herbivores produce faster-decomposing litter (Grime et al. 1996; Schädler et al. 2003; Pálková and Leps 2008; Kurokawa et al. 2010). Therefore, it is hypothesized that selective feeding by herbivores results in slow decomposition of leaf litter, according to such a positive relationship between palatability and decomposability (Hartley and Jones 2004; Bardgett and Wardle 2010). That is, highly palatable (= highly decomposable) plants are selectively consumed by herbivores, subsequently poorly palatable (= poorly decomposable) plants remain, and therefore, overall litter decomposition will be slow. However, a few studies revealed that there are no positive relationships between leaf palatability and litter decomposability in tropical and temperate forests (Kurokawa and Nakashizuka 2008; Silva and Vasconcelos 2011; Kagata and Ohgushi 2011), indicating that a positive correlation between palatability and decomposability is not always true in a wide range of plant-insect systems. The lack of a positive relationship suggests that some differences in key factors determine palatability and decomposability. In general, litter decomposability is largely dependent on the litter quality, e.g. concentrations of nitrogen, phosphorus, tannins, and lignin (Gallardo and Merino 1993; Aerts 1997; Kraus et al. 2003), although soil microclimate and soil organisms are also important in determining the decomposition process (Seastedt 1984; Aerts 1997). In particular, tannins and lignin are critical chemicals that suppress litter decomposition (Kraus et al. 2003). These chemicals also have a function as defensive chemicals against insect herbivores (Feeny 1970), and they result in a positive relationship between palatability and decomposability (Bardgett and Wardle 2010). However, plants display various types of defenses against herbivores (see below), and those defenses often co-occur in individual plants of the same species or different tissues of an individual (Koricheva et al. 2004). Hence, the palatability-decomposability relationship would display positive, negative, or no correlation, depending on the types of antiherbivore defenses in plants.

2.3.2   Diversity of Antiherbivore Defense in Plants

Antiherbivore defenses in plants are primarily categorized into three types: physical, chemical, and biotic defenses. Typical traits of physical defense include tough leaves, trichomes, and spines, which generally deter herbivore feeding (Schoonhoven et al. 1998). Chemical defenses employ plant secondary metabolites, and they are further divided to quantitative and qualitative defenses (Coley et al. 1985).

The quantitative defenses are characterized by high molecular weight compounds, such as polyphenoles, tannins, and lignin (cf. carbon-based defense compounds), which generally inhibit the digestion of herbivores (Harborne 1993). The qualitative defenses are characterized by low molecular weight compounds, such as alkaloids, glucosinolates, and cyanogenic glycosides (cf. nitrogen-based defense compounds), which are strongly toxic (Harborne 1993). While the physical and chemical defenses are direct defense systems by the plant itself, biotic defenses are indirect defenses that operate by attracting natural enemies of insect herbivores (Heil 2008). There is another classification for plant antiherbivore defenses, i.e., constitutive and induced defenses. A constitutive defense is a defense strategy in which a high primary defense level is maintained irrespective of the presence or absence of herbivores, whereas an induced defense is a strategy in which the defense level is usually low but is rapidly increased in response to herbivory (Schoonhoven et al. 1998). Furthermore, resistance and tolerance are other classifications for plant antiherbivore strategy. Resistance is a strategy to avoid or reduce the amount of herbivory damage, and tolerance is a strategy to compensate for the negative fitness effect caused by herbivory damage, for example, by compensatory regrowth (Núñez-Farfán et al. 2007).

Since the maintenance of multiple defense systems would be costly for a plant, several researchers have suggested the existence of constraints on simultaneous resource allocation to defense strategies, resulting in trade-offs between different types of defenses (Koricheva et al. 2004). Such a concept of trade-off between antiherbivore defenses is the basis of most of the plant defense theories, such as the carbon and nutrient balance hypothesis (Bryant et al. 1983), the resource availability hypothesis (Coley et al. 1985), and the growth-differentiation balance hypothesis (Herms and Mattson 1992). Several studies have found a trade-off between types of antiherbivore defenses, for example, between physical and chemical defenses (Twigg and Socha 1996), between chemical and biotic defenses (Eck et al. 2001), and between qualitative and quantitative defenses (Stevens et al. 1995). However, others did not detect trade-offs between different types of antiherbivore defenses (e.g. Steward and Keeler 1988; Agrawal and Fishbein 2006). Meta-analyses also examined whether there is a negative association between types of antiherbivore defenses (Koricheva et al. 2004; Leimu and Koricheva 2006). They showed that significant negative correlation was only detected between constitutive and induced defenses, but no significant correlation was detected between the other types of defenses, such as qualitative and quantitative defenses. Therefore, plant defense theories that are based on a trade-off remain a controversial issue.

2.3.3 Relationship Between Antiherbivore Defenses and Litter Decomposability

Antiherbivore defenses of plants determine not only leaf palatability to insect herbivores but also litter decomposability. This is because characteristics of living leaves are generally taken over to abscised leaves (e.g. Kurokawa and Nakashizuka 2008). However, the relationship between types of antiherbivore defenses and litter decomposability has been poorly explored except regarding a few defensive strategies. For example, quantitative chemical defense compounds, such as tannins, are well known to decelerate litter decomposition, because these compounds generally suppress the activity of microbial decomposers (Kraus et al. 2003). In addition, litter from the plants that have regrown in response to herbivory is likely to be decomposed rapidly due to increased quality of the litter (Hunter 2001). The effects of other defensive strategies, such as qualitative chemical defense, on litter decomposition are poorly explored (but see Siegrist et al. 2010). Therefore, studies on the relationship between types of antiherbivore defenses in plants and litter decomposability are needed to obtain a general picture about the relationship between leaf palatability and litter decomposability.

2.3.4   Intraspecific Variation in Palatability and Decomposability

Most studies that examined the relationship between leaf palatability and litter decomposability have focused on interspecific variation of plants in these two parameters (Grime et al. 1996; Schädler et al. 2003; Pálková and Leps 2008; Kurokawa et al. 2010). On the other hand, a large number of studies have demonstrated that the level of antiherbivore defenses and subsequent insect herbivory were variable, depending on the phenotype and genotype of a plant (e.g. Osier et al. 2000). Moreover, there is accumulating evidence that litter decomposition differs among plant genotypes (Schweitzer et al. 2005a; Silfver et al. 2007; Crutsinger et al. 2009). These studies indicate that intraspecific genetic and phenotypic variation of plants is important in driving ecosystem properties (Whitham et al. 2006; Post and Palkovacs 2011). However, few studies have explicitly examined the relationship between palatability and decomposability from the point of view of intraspecific variation in plants. To date, two series of such studies have revealed a relationship between leaf palatability and litter decomposability in this context. One examined a poplar-beaver system (e.g. Whitham et al. 2006), and the other concerned a pine-insect herbivores system (e.g. Chapman et al. 2003; Classen et al. 2007); (1) Poplar-beaver system—Concentration of condensed tannin in bark and leaves differed among genotypes of poplar trees (Driebe and Whitham 2000). Beavers avoided harvesting the poplar branches with higher levels of condensed tannin (Bailey et al. 2004), and this behavior left a larger proportion of poplar trees with higher levels of condensed tannin (Whitham et al. 2006). Litter from poplar trees with rich condensed tannin showed slower decomposition and nitrogen mineralization (Whitham et al. 2006), indicating a positive correlation between palatability to beavers and litter decomposability through intraspecific variation in condensed tannin levels. Therefore, beavers are likely to decelerate nutrient cycling in an ecosystem dominated by poplar trees, and poplar genotypes with higher levels of condensed tannin may be selected to display adaptations to cope with poor nutrients due to the decelerated nutrient cycling (Post and Palkovacs 2011). (2) Pine-insect herbivores system—There are distinct resistant and susceptible trees to pine needle scale or stem-boring moth in piñion pine, and these traits are genetically determined (Whitham and Mopper 1985; Cobb and Whitham 1993). Nitrogen concentration in needles differed, but lignin and tannins concentration did not differ, between the resistant and susceptible trees (Chapman et al. 2003). Needle litter decomposition was not affected by herbivore-susceptibility (Classen et al. 2007). This indicates that there is no positive correlation between needle palatability to the two insect herbivores and litter decomposability. On the other hand, insect herbivory induced premature needle abscission, resulting in accelerated litter decomposition and nutrient release (Chapman et al. 2003; Chapman 2006). These finding imply that genetically based needle traits that are associated with palatability are less important, and instead phenotypic plasticity of the traits induced by insect herbivory is more important, in determining the litter decomposition.

Thus, these examples clearly showed that the relationship between palatability and decomposability is not always a positive correlation in terms of intraspecific genetic variation, as well as interspecific variation, and that herbivore can be a driver in determining plant characteristics that are based on genetic and phenotypic levels, which affect litter decomposition and nutrient cycling.

In terrestrial systems, it is generally accepted that insect herbivores have a relatively small impact on the fraction of energy and nutrients inputs in the decomposition processes because of the low herbivory rate, i.e., less than 20 % (Cyr and Pace 1993; Cebrian and Lartigue 2004). However, various insect herbivores can potentially undergo outbreaks in which they reach extremely high density (Kamata 2002), as a result of which the insect herbivory and excretion of waste products have a critical effect on the decomposition process and nutrient cycling (Hunter 2001; Lovett et al. 2002; Clark et al. 2010). Furthermore, chronic insect herbivory would be one of the important selective forces in the evolution of antiherbivore defenses in plants (Whitham and Mopper 1985). This is supported by evidence that invading plants, which escape from native insect herbivores, lower their levels of antiherbivore defenses (Beaton et al. 2011). If insect herbivores function as a significant selective force for the evolution plant characteristics, they have the potential to affect ecosystem processes through evolutionary changes in plant characteristics, as well as through phenotypic changes in plant characteristics (Fig. 2.2). Furthermore, changes in the ecosystem processes may feedback to the plant characteristics because the level and type of antiherbivore defenses can be influenced by soil nutrient status (Bryant et al. 1983). The feedback through litter decomposition may be important in plants with poorly palatable leaves because loss of leaf mass due to herbivory is expected to be small. On the other hand, the feedback through insect excrement may become more important in plants with highly palatable leaves because a larger amount of the insect excrements is expected. In addition, the recent development of molecular techniques will reveal the genetic basis of evolution of antiherbivore defensive traits in plants (Anderson and Mitchell-Olds 2011). Coupled with knowledge of such a genetic-base of plant antiherbivore defenses, an understanding of the relationship between leaf palatability and litter decomposability in the context of intraspecific variation would clarify the roles of insect herbivores in ecosystem processes through the evolution of antiherbivore defenses in plants.

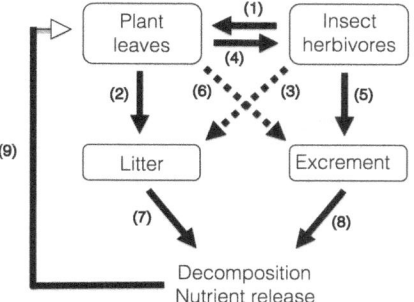

Fig. 2.2 Ecosystem consequences of insect–plant interactions. (*1*) Insect herbivores can change plant characteristics in both genetic and phenotypic levels. (*2*) Characteristics of living leaves are carried over to abscised leaves, and therefore, (*3*) insect herbivores indirectly alter litter inputs, in terms of quality and quantity, to a decomposition system. On the other hand, (*4*) plant characteristics determine the leaf palatability to insect herbivores. (*5*) Insect herbivores excrete the waste produced as consequence of leaf herbivory. (*6*) Plants indirectly determine inputs of insect excrement to a decomposition system, since quality and quantity of insect excrement are largely determined by leaf characteristics. (*7*) and (*8*) Quality and quantity of leaf litter and insect excrements influence their decomposition processes, and subsequently nutrient release to the soil. (*9*) Soil nutrient status affects plant characteristics, and therefore, the effects of the insect-plant interaction feedback to plant characteristics through ecosystem processes

2.4   Future Directions

Our review clearly illustrated that genetically controlled and plastically modified characteristics of plants are undoubtedly important to determine community and ecosystem properties. In particular, most of such plant traits that influence community and ecosystem properties are related to antiherbivore defense. Therefore, elucidating the evolutionary processes of antiherbivore defense of plants would be an effective approach to the integration of evolutionary ecology with community and ecosystem ecology. Phenotypes of antiherbivore defense of plants would be largely influenced by plant genotype, type and level of herbivory, and soil nutrient availability (Bryant et al. 1983; Agrawal et al. 2002) (Fig. 2.3). Herbivore-mediated processes that affect defensive traits are associated with diffuse co-evolution of plant(s) and herbivore(s) (Janzen 1980), and soil-mediated processes are associated with plant-soil feedback (Ehrenfeld et al. 2005). Although each of these processes (diffuse co-evolution process and plant-soil feedback process) has yet to be rigorously examined, integrating the two processes is a challenging and novel approach to understanding the evolution of antiherbivore defenses of plants. For example, the following questions are important to understand the evolution of antiherbivore defensive traits of plants in the community and ecosystem context: (1) How does community structure of herbivores influence ecosystem processes? (2) Which selective pressures from the diffuse co-evolution process and the plant-soil feedback process are relatively strong in their effects on the evolution of plant defense traits? (3) Do the selective pressures from the two processes result in the same, opposite,

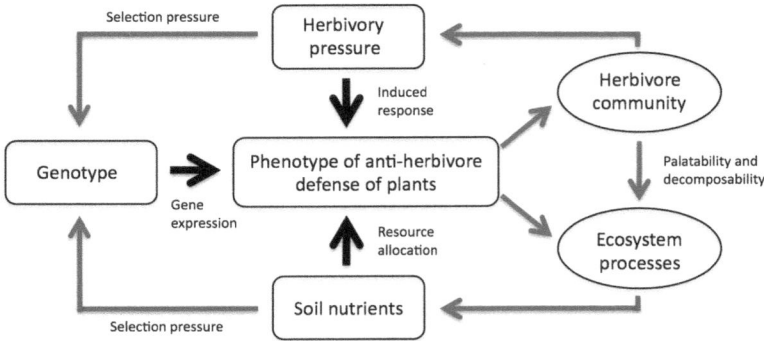

Fig 2.3 Evolution of anti-herbivore defensive traits of plants in community and ecosystem context

or independent directions of evolution of plant defensive traits? (4) How is phenotypic plasticity of plant characteristics in response to herbivory and nutrient availability important in the diffuse co-evolutionary process and plant-soil feedback process? In particular, the last question would provide novel insights about community/ecosystem genetics that would clarify the evolutionary process of plant traits. Numerous studies have demonstrated that the phenotype of antiherbivore defensive traits of plants is strongly influenced by herbivory and soil nutrient availability, without genotypic alteration (e.g. Karban and Baldwin 1997; Kim et al. 2002). Such phenotypic plasticity in plant defenses undoubtedly influences community structures of plant-associated arthropods and ecosystem processes such as decomposition rate and nutrient release (Nakamura et al. 2003; Schweitzer et al. 2005b). Therefore, not only genetically controlled defensive traits of plants but also plastically modified defensive traits should be taken into consideration to understand the evolution of antiherbivore defensive traits of plants in community and ecosystem contexts.

References

Aerts R (1997) Climate, leaf litter chemistry and leaf litter decomposition in terrestrial ecosystems: a triangular relationship. Oikos 79:439–449

Agrawal AA (1998) Induced responses to herbivory and increased plant performance. Science 279:1201

Agrawal AA (2000) Specificity of induced resistance in wild radish: causes and consequences for two specialist and two generalist caterpillars. Oikos 89:493–500

Agrawal AA (2001) Phenotypic plasticity in the interactions and evolution of species. Science 294:321

Agrawal AA (2006) Macroevolution of plant defense strategies. Trends Ecol Evol 22:103–109

Agrawal AA (2011) Current trends in the evolutionary ecology of plant defense. Funct Ecol 25:420–432

Agrawal AA, Fishbein M (2006) Plant defense syndromes. Ecology 87:S132–S149

Agrawal AA, Strauss SY, Stout MJ (1999) Costs of induced responses and tolerance to herbivory in male and female fitness components of wild radish. Evolution 53:1093–1104

Agrawal AA, Conner JK, Johnson MTJ, Wallsgrove R (2002) Ecological genetics of an induced plant defense against herbivores: additive genetic variance and costs of phenotypic plasticity. Evolution 56:2206–2213

Alborn H, Turlings T, Jones T, Stenhagen G, Loughrin J, Tumlinson J (1997) An elicitor of plant volatiles from beet armyworm oral secretion. Science 276:945

Anderson JT, Mitchell-Olds T (2011) Ecological genetics and genomics of plant defenses: evidence and approaches. Funct Ecol 25:312–324

Ando Y, Ohgushi T (2008) Ant-and plant-mediated indirect effects induced by aphid colonization on herbivorous insects on tall goldenrod. Popul Ecol 50:181–189

Ando Y, Utsumi S, Ohgushi T (2011a) Community-wide impact of an exotic aphid on introduced tall goldenrod. Ecol Entomol 36:643–653

Ando Y, Utsumi S, Craig TP, Itami J, Ohgushi T (2011b) How are arthropod communities organized on an introduced plant *Solidago altissima*? J Plant Interac 6:169–170

Bailey JK, Schweitzer JA, Rehill BJ, Lindroth RL, Martinsen GD, Whitham TG (2004) Beavers as molecular genetics: a genetic basis to the foraging of an ecosysten engineer. Ecology 85:603–608

Bardgett RD, Wardle DA (2010) Aboveground-belowground linkages. Oxford University Press, New York

Beaton LL, van Zandt PA, Esselman EJ, Knight TM (2011) Comparison of the herbivore defense and competitive ability of ancestral and modern genotypes of an invasive plant, *Lespedeza cuneata*. Oikos 120:1413–1419

Bingham RA, Agrawal AA (2010) Specificity and trade-offs in the induced plant defence of common milkweed *Asclepias syriaca* to two lepidopteran herbivores. J Ecol 98:1014–1022

Bryant JP, Chapin FS, Klein B (1983) Carbon/nutrient balance of boreal plants in relation to vertibrate herbivory. Oikos 40:357–368

Cebrian J, Lartigue J (2004) Patterns of herbivory and decomposition in aquatic and terrestrial ecosystems. Ecol Monogr 74:237–259

Chapman SK (2006) Herbivory differentially alters plant litter dynamics of evergreen and deciduous trees. Oikos 114:566–574

Chapman SK, Hart SC, Cobb NS, Whitham TG, Koch GW (2003) Insect herbivory increases litter quality and decomposition: an extension of the acceleration hypothesis. Ecology 84:2867–2876

Clark KL, Skowronski N, Hom J (2010) Invasive insects impact forest carbon dynamics. Glob Chan Biol 16:88–101

Classen AT, Chapman SK, Whitham TG, Hart SC, Koch GW (2007) Genetic-based plant resistance and susceptibility traits to herbivory influence needle and root litter nutrient dynamics. J Ecol 95:1181–1194

Cobb NS, Whitham TG (1993) Herbivore deme formation on individual trees: a test case. Oecologia 94:496–502

Coley PD, Bryant JP, Chapin FS (1985) Resource availability and plant antiherbivore defense. Science 230:895–899

Crutsinger GM, Sanders NJ, Classen AT (2009) Comparing intra- and inter-specific effects on litter decomposition in an old-field ecosystem. Basic Appl Ecol 10:535–543

Cyr H, Pace ML (1993) Magnitude and patterns of herbivory in aquatic and terrestrial ecosystems. Nature 361:148–150

DeWitt TJ, Scheiner SM (2004) Phenotypic plasticity. Oxford University Press, Oxford

Driebe EM, Whitham TG (2000) Cottonwood hybridization affects tannin and nitrogen content of leaf litter and alters decomposition. Oecologia 123:99–107

Eck G, Fiala B, Linsenmair KE, Hashim RB, Proksch P (2001) Trade-off between chemical and biotic antiherbivore defense in the south east Asian plant genus *Macaranga*. J Chem Ecol 27:1979–1996

Ehrenfeld JG, Ravit B, Elgersma K (2005) Feedback in the plant-soil system. Annu Rev Environ Res 30:75–115

English-Loeb G, Karban R, Walker MA (1998) Genotypic variation in constitutive and induced resistance in grapes against spider mite (Acari: Tetranychidae) herbivores. Environ Entomol 27:297–304

Feeny P (1970) Seasonal changes in oak leaf tannins and nutrients as a cause of spring feeding by winter moth caterpillars. Ecology 51:555–581

Gallardo A, Merino J (1993) Leaf decomposition in two mediterranean ecosystems of southwest Spain: influence of substrate quality. Ecology 74:152–161

Geervliet JBF, Posthumus MA, Vet LEM, Dicke M (1997) Comparative analysis of headspace volatiles from different caterpillar-infested or uninfested food plants of *Pieris* species. J Chem Ecol 23:2935–2954

Grime JP, Cornelissen JH, Hodgson JG (1996) Evidence of a causal connection between anti-herbivore defence and the decomposition rate of leaves. Oikos 77:489–494

Harborne JB (1993) Introduction to ecological biochemistry. Academic, London

Hartley SE, Jones TH (2004) Insect herbivores, nutrient cycling and plant productivity. In: Weisser WW, Siemann E (eds) Insects and ecosystem function. Springer, Berlin, pp 27–52

Hartley S, Lawton J (1987) Effects of different types of damage on the chemistry of birch foliage, and the responses of birch feeding insects. Oecologia 74:432–437

Heil M (2008) Indirect defence via tritrophic interactions. New Phytol 178:41–61

Herms DA, Mattson WJ (1992) The dilemma of plants: to grow or defend. Q Rev Biol 67:283–335

Hochwender CG, Janson EM, Cha DH, Fritz RS (2005) Community structure of insect herbivores in a hybrid system: examining the effects of browsing damage and plant genetic variation. Ecol Entomol 30:170–175

Hunter MD (2001) Insect population dynamics meets ecosystem ecology: effects of herbivory on soil nutrient dynamics. Agric For Entomol 3:77–84

Janzen DH (1980) When is it coevolution. Evolution 34:611–612

Juenger T, Bergelson J (2000) The evolution of compensation to herbivory in scarlet gilia, *Ipomopsis aggregata*: herbivore-imposed natural selection and the quantitative genetics of tolerance. Evolution 54:764–777

Kagata H, Ohgushi T (2011) Ecosystem consequences of selective feeding of an insect herbivore: palatability-decomposability relationship revisited. Ecol Entomol 36:768–775

Kagata H, Ohgushi T (2012) Positive and negative impacts of insect frass quality on soil nitrogen availability and plant growth. Popul Ecol 54:75–82

Kamata N (2002) Outbreaks of forest defoliating insects in Japan, 1950–2000. Bull Entomol Res 92:109–117

Karban R, Baldwin IT (1997) Induced responses to herbivory. University of Chicago Press, Chicago

Kay AD, Mankowski J, Hobbie SH (2008) Long-term burning interacts with herbivory to slow decomposition. Ecology 89:1188–1194

Kim S, Matuso T, Watanabe M, Watanabe Y (2002) Effect of nitrogen and sulphur application on the glucosinolate content in vegetable turnip rape (*Brassica rapa* L.). Soil Sci Plant Nutrit 48:43–49

Koricheva J, Nykänen H, Gianoli E (2004) Meta-analysis of trade-offs among plant antiherbivore defenses: are plants jacks-of-all-trades, masters of all? Am Nat 163:E65–E75

Korth KL, Dixon RA (1997) Evidence for chewing insect-specific molecular events distinct from a general wound response in leaves. Plant Physiol 115:1299

Kraus TEC, Dahlgren RA, Zasoski RJ (2003) Tannins in nutrient dynamics of forest ecosystems - a review. Plant Soil 256:41–66

Krause SC, Raffa KF (1992) Comparison of insect, fungal, and mechanically induced defoliation of larch: effects on plant productivity and subsequent host susceptibility. Oecologia 90:411–416

Kurokawa H, Nakashizuka T (2008) Leaf herbivory and decomposability in a Malaysian tropical rain forest. Ecology 89:2645–2656

Kurokawa H, Peltzer DA, Wardle DA (2010) Plant traits, leaf palatability and litter decomposability for co-occurring woody species differing in invasion status and nitrogen fixation ability. Funct Ecol 24:513–523

Leimu R, Koricheva J (2006) A meta-analysis of tradeoffs between plant tolerance and resistance to herbivores: combining evidence from ecological and agricultural studies. Oikos 112:1–9

Lin H, Kogan M, Fischer D (1990) Induced resistance in soybean to the Mexican bean beetle (Coleoptera: Coccinellidae): comparisons of inducing factors. Environ Entomol 19: 1852–1857

Lovett GM, Christenson LM, Groffman PM, Jones CG, Hart JE, Mitchell MJ (2002) Insect defoliation and nitrogen cycling in forests. Bioscience 52:335–341

McAuslane HJ, Alborn HT (1998) Systemic induction of allelochemicals in glanded and glandless isogenic cotton by *Spodoptera exigua* feeding. J Chem Ecol 24:399–416

McCloud ES, Baldwin IT (1997) Herbivory and caterpillar regurgitants amplify the wound-induced increases in jasmonic acid but not nicotine in Nicotiana sylvestris. Planta 203:430–435

McNaughton S (1983) Compensatory plant growth as a response to herbivory. Oikos 40:329–336

Miner BG, Sultan SE, Morgan SG, Padilla DK, Relyea RA (2005) Ecological consequences of phenotypic plasticity. Trend Ecol Evol 20:685–692

Moon AM, Dyer M, Brown M, Crossley D (1994) Epidermal growth factor interacts with indole-3-acetic acid and promotes coleoptile growth. Plant Cell Physiol 35:1173

Nakamura M, Miyamoto Y, Ohgushi T (2003) Gall initiation enhances the availability of food resources for herbivorous insects. Funct Ecol 17:851–857

Núñez-Farfán J, Fornoni J, Valverde PL (2007) The evolution of resistance and tolerance to herbivores. Annu Rev Ecol Evol Syst 38:541–566

Ohgushi T (2005) Indirect interaction webs: herbivore-induced effects through trait change in plants. Annu Rev Ecol Evol Syst 36:81–105

Ohgushi T, Ando Y, Utsumi S, Craig TP (2011) Indirect interaction webs on tall goldenrod: community consequences of herbivore-induced phenotypes and genetic variation of plants. J Plant Interac 6:147–150

Osier TL, Hwang S, Lindroth RL (2000) Effects of phytochemical variation in quaking aspen *Populus tremuloides* clones on gypsy moth *Lymantria dispar* performance in the field and laboratory. Ecol Entomol 25:197–207

Pálková K, Leps J (2008) Positive relationship between plant palatability and litter decomposition in meadow plants. Comm Ecol 9:17–27

Poelman EH, Van L, Joop J, Van Nicole M, Vet LEM, Dicke M (2010) Herbivore-induced plant responses in *Brassica oleracea* prevail over effects of constitutive resistance and result in enhanced herbivore attack. Ecol Entomol 35:240–247

Post DM, Palkovacs EP (2011) Eco-evolutionary feedbacks in community and ecosystem ecology: interactions between the ecological theatre and the evolutionary play. Phi Trans Royal Soc B 364:1629–1640

Power AG (1988) Leafhopper response to genetically diverse maize stands. Entomol Exp Appl 49:213–219

Rosenthal J, Kotanen P (1994) Terrestrial plant tolerance to herbivory. Trend Ecol Evol 9: 145–148

Schädler M, Jung G, Auge H, Brandl R (2003) Palatability, decomposition and insect herbivory: patterns in a successional old-field plant community. Oikos 103:121–132

Schoonhoven LM, Jermy T, van Loon JJA (1998) Insect-plant biology. Chapman & Hall, London

Schweitzer JA, Bailey JK, Hart SC, Whitham TG (2005a) Nonadditive effects of mixing cottonwood genotypes on litter decomposition and nutrient dynamics. Ecology 86:2384–2840

Schweitzer JA, Bailey JK, Hart SC, Wimp GM, Chapman SK, Whitham TG (2005b) The interaction of plant genotype and herbivory decelerate leaf litter decomposition and alter nutrient dynamics. Oikos 110:133–145

Seastedt TR (1984) The role of microarthropods in decomposition and mineralization processes. Annu Rev Entomol 29:25–46

Siegrist JA, McCulley RL, Bush LP, Phillips TD (2010) Alkaloids may not be responsible for endophyte-associated reductions in tall fescue decomposition rates. Funct Ecol 24:460–468

Silfver T, Mikola J, Rousi M, Roininen H, Oksanen E (2007) Leaf litter decomposition differs among genotypes in a local *Betula pendula* population. Oecologia 152:707–714

Silva LVB, Vasconcelos HL (2011) Plant palatability to leaf-cutter ants (*Atta laevigata*) and litter decomposability in a Neotropical woodland savanna. Aust Ecol 36:504–510

Simms EL (1992) Costs of plant resistance to herbivores. In: Fritz RS, Simms EL (eds) Plant resistance to herbivores and pathogens. Ecology, evolution, and genetics. University of Chicago Press, Chicago, pp 392–425

Snoeren TAL, Kappers IF, Broekgaarden C, Mumm R, Dicke M, Bouwmeester HJ (2010) Natural variation in herbivore-induced volatiles in *Arabidopsis thaliana*. J Exp Bot 61:3041–3056

Stadler B, Mühlenberg E, Michalzik B (2004) The Ecology driving nutrient fluxes in forests. In: Weisser WW, Siemann E (eds) Insects and ecosystem function. Springer, Berlin, pp 213–239

Stevens JF, Hart HT, van Ham RCHJ, Elema ET, van den Ent MMVX, Wildeboer M, Zwaving JH (1995) Distribution of alkaloids and tannins in the Crassulaceae. Biochem Syst Ecol 23: 157–165

Steward JL, Keeler KH (1988) Are there trade-offs among antiherbivore defenses in Ipomoea (Convolvulaceae)? Oikos 53:79–86

Stout MJ, Brovont RA, Duffey SS (1998) Effect of nitrogen availability on expression of constitutive and inducible chemical defenses in tomato, *Lycopersicon esculentum*. J Chem Ecol 24:945–963

Stowe KA, Marquis RJ, Hochwender CG, Simms EL (2000) The evolutionary ecology of tolerance to consumer damage. Annu Rev Ecol Syst 31:565–595

Strauss SY, Agrawal AA (1999) The ecology and evolution of plant tolerance to herbivory. Trends Ecol Evol 14:179–185

Strauss SY, Sahli H, Conner JK (2005) Toward a more trait-centered approach to diffuse (co)evolution. New Phytol 165:81–89

Tétard-Jones C, Kertesz MA, Gallois P, Preziosi RF (2007) Genotype-by-genotype interactions modified by a third species in a plant-insect system. Am Nat 170:492–499

Thompson JN (2005) The geographic mosaic of coevolution. University of Chicago Press, Chicago

Turlings T, Loughrin JH, McCall PJ, Röse U, Lewis WJ, Tumlinson JH (1995) How caterpillar-damaged plants protect themselves by attracting parasitic wasps. Proc Nat Acad Sci USA 92:4169

Twigg LE, Socha LV (1996) Physical versus chemical defence mechanisms in toxic Gasrolobium. Oecologia 108:21–28

Underwood N, Morris W, Gross K, Lockwood JR (2000) Induced resistance to Mexican bean beetles in soybean: variation among genotypes and lack of correlation with constitutive resistance. Oecologia 122:83–89

Utsumi S (2011) Eco-evolutionary dynamics in herbivorous insect communities mediated by induced plant responses. Popul Ecol 53:23–34

Utsumi S, Ohgushi T (2007) Plant regrowth response to a stem-boring insect: a swift moth-willow system. Popul Ecol 49:241–248

Utsumi S, Ohgushi T (2009) Community-wide impacts of herbivore-induced plant regrowth on arthropods in a multi-willow species system. Oikos 118:1805–1815

Utsumi S, Nakamura M, Ohgushi T (2009) Community consequences of herbivore-induced bottom-up trophic cascades: the importance of resource heterogeneity. J Anim Ecol 78:953–963

Utsumi S, Ando Y, Miki T (2010) Linkages among trait-mediated indirect effects: a new framework for the indirect interaction web. Popul Ecol 58:1–13

Utsumi S, Ando Y, Craig TP, Ohgushi T (2011) Plant genotypic diversity increases population size of a herbivorous insect. Proc Royal Soc B 278:3108–3115

van Dam NM, Vrieling K (1994) Genetic variation in constitutive and inducible pyrrolizidine alkaloid levels in *Cynoglossum officinale* L. Oecologia 99:374–378

Van Zandt PA, Agrawal AA (2004) Community-wide impacts of herbivore-induced plant responses in milkweed (*Asclepias syriaca*). Ecology 85:2616–2629

Whitham TG, Mopper S (1985) Chronic herbivory: impacts on architecture and sex expression of pinyon pine. Science 228:1089–1091

Whitham TG, Bailey JK, Schweitzer JA, Shuster SM, Bangert RK, LeRoy CJ, Lonsdorf EV, Allan GJ, DiFazio SP, Potts BM, Fischer DG, Gehring CA, Lindroth RL, Marks JC, Hart SC, Wimp GM, Wooley SC (2006) A framework for community and ecosystem genetics: from genes to ecosystems. Nat Rev Genet 7:510–523

Zangerl AR, Berenbaum MR (1990) Furanocoumarin induction in wild parsnip: genetics and population variation. Ecology 71:1933–1940

Chapter 3
Accelerated Diversification by Spatial and Temporal Isolation Associated with Life-History Evolution in Insects

Abstract Life history evolution may alter the mode and rate of the speciation of a lineage. Recent studies on diversification processes associated with life history evolution in insects are introduced here. First, the hypothesis that the repeated evolution of flightlessness in winged insects may have promoted speciation through reduced gene flow among local populations was tested for one of the largest group of insects, the Coleoptera. A close examination of the geographic differentiation and speciation rates in carrion beetles (Silphinae) in relation to flight capability generally supported the hypothesis and suggested that evolution of flightlessness contributed to beetle diversity. Second, an allochronic speciation hypothesis was tested for a geometrid winter moth for which the reproductive period between early and late winter is disrupted by severe midwinter conditions at high latitudes or elevations. The formation of genetically divergent allochronic populations was widely and repeatedly observed in habitats with severe winters, but not in habitats with mild winters. This suggests that seasonal climatic harshness can act as a temporal barrier leading to reproductive isolation and is a potentially important driving force for allochronic speciation in insects. These novel diversification processes provide new insights into species diversity and the speciation processes of insects in spatially heterogeneous and seasonally variable environments.

Keywords Adaptation • Allochronic speciation • Allopatric speciation • Dispersal • Flightlessness • Insect flight • Speciation

3.1 Introduction

The understanding of speciation processes has recently been broadened by abundant evidence for several different types of ecological speciation in the wild (e.g., Coyne and Orr 2004; Dieckmann et al. 2004; Rundle and Nosil 2005; Sobel et al. 2009). The role of ecological divergence in speciation is most clearly demonstrated in cases of sympatric or parapatric population differentiation (Coyne and Orr 2004). Nonetheless, allopatric speciation processes may still have played a major role. One aspect of allopatric speciation that has not been well explored is how it is connected to the evolution of dispersal, which itself is highly relevant to the ecology of speciation. The evolution of dispersal has been discussed theoretically (e.g., Southwood 1977), but the relationship between the evolution of dispersal ability and speciation rate has not been well explored, despite the simple and intuitive prediction that a low dispersal rate would promote allopatric speciation. Here, we introduce our recent study of the relationship between the evolution of flightlessness and diversification in beetles, and demonstrate how the loss of flight can promote beetle diversification (Ikeda et al. 2012).

Unlike allopatric speciation, allochronic speciation (i.e., speciation due to temporal isolation) has been considered to be much more difficult and thus rarer in nature. However, insects in seasonal environments must stop development and reproduction during the period of climatic harshness (such as a cold winter or dry and hot summer). The period of climatic harshness can act as a barrier for gene flow when it disrupts reproductive seasons. The resultant temporal isolation may facilitate speciation between allochronic populations. This type of speciation was first proposed by Alexander and Bigelow (1960), but convincing evidence has not yet been provided. Here, we introduce our recent study on incipient allochronic speciation by climatic disruption of the reproductive period for a winter moth lineage (Yamamoto and Sota 2009, 2012). The two studies reviewed here have been conducted with my colleagues during our term with the Global Center of Excellence Program: Evolution and Biodiversity at Kyoto University.

3.2 Evolution of Flightlessness May Promote Allopatric Speciation

Dispersal is the dominant life history strategy of organisms for coping with fluctuating habitat conditions in space and time (Southwood 1977). Flight is the most powerful method of dispersal, but among modern biota, only insects, birds, and bats possess spontaneous flight ability. Insects are thought to have developed wings for flight as early as 400 million years ago, and this evolutionary event could have promoted the diversification of insects by facilitating the colonization of a variety regions and habitats (Roff 1990; Engel and Grimaldi 2004; Grimaldi and Engel 2005). Flight ability enables individuals to find mates and food over wider areas and assists with colonizing distant habitats, thus increasing the chance of reproductive success (Roff 1986; Roff and Fairbairn 1991). Despite the benefits of flight, however, the loss of flight ability has occurred repeatedly in various insect groups (Roff 1990, 1994; Wagner and Liebherr 1992). The loss of flight ability in winged insects can occur for several reasons, but a key factor in the evolution of flightlessness is the high cost of flight in terms of resource investment, both structural and energetic (Roff 1991; Lövei and Sunderland 1996; Zera and Denno 1997). This results in a reduction of

Fig. 3.1 Dispersal ability
and genetic differentiation
of local populations

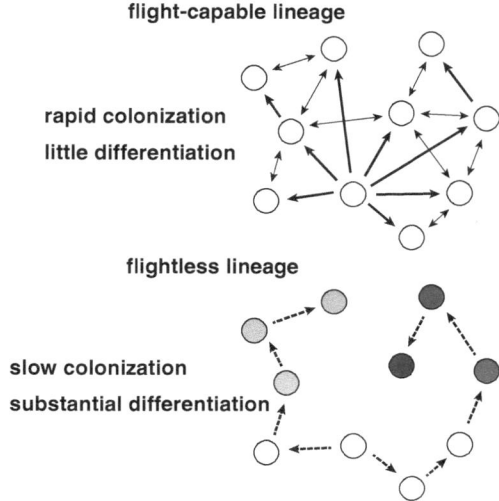

resource allocation to reproduction and survival, and mortality increases during dispersal by flight. The incidence of flightlessness in each insect lineage differs with habitat type and generally tends to increase in stable habitats. However, generalizing the relationship between habitat stability and the incidence of flightlessness is difficult because habitat stability is difficult to define (Roff 1990). Under persistently favorable habitat conditions (i.e., in stable habitats), selection for flight ability may be weakened by the decreased necessity for spatial dispersal, and the loss of flight muscles or wings may occur due to the mutation of genes involved in the formation of the flight apparatus. However, selection pressures for maintaining flight ability may still exist under stable habitat conditions, e.g., for finding mates (especially of non-kin) and resources of temporal and sporadic availability.

Dispersal promotes gene flow among local populations and tends to prevent local populations from diverging genetically. Therefore, limitation of dispersal in either a passive or spontaneous way should allow local populations to diverge (e.g., Zera 1981; Smith and Farrell 2006). In light of this general prediction, one may expect that the reduced dispersal ability among winged insects due to the loss of flight ability promotes genetic differentiation among geographic populations (Fig. 3.1), which eventually results in allopatric speciation. Despite the potentially intimate connection between the evolution of flight/flightlessness and the speciation rate of a lineage, this relationship has not been examined in depth in empirical and theoretical studies. We can expect that among related groups of insects, flightless lineages have attained higher speciation rates, or possess higher potential for allopatric speciation, than flight-capable lineages. Although the idea that lower levels of dispersal may result in higher rates of allopatric speciation may not be novel (e.g., Hansen 1983; Jablonski 1986), empirical examination of the consequences of flightlessness in the evolution of insect species diversity is of great potential importance.

3.2.1   Loss of Flight Promotes Beetle Diversification: Hypothesis Testing with Carrion Beetles

Beetles (Coleoptera) are the most species-rich group of insects, with 0.35 million known species comprising about 40 % of known insect species (Grimaldi and Engel 2005). This high species richness has attracted the attention of biologists, and adaptive radiation in angiosperm-associated beetles has been considered the major factor driving this diversity (Farrell 1998). However, a recent comprehensive analysis of beetle phylogeny revealed that most of the extant groups originated before the Jurassic (i.e., preceding the radiation of angiosperms) and have continued to diversify (Hunt et al. 2007). In addition, no significant differences in speciation rates exist between those groups associated and not associated with angiosperms, contrary to Farrell's (1998) assessment (Hunt et al. 2007). The secrets of beetle diversification may stem from mechanistic aspects based on the particular design of beetle body (Grimaldi and Engel 2005). The beetle body is covered with a hard prothorax and elytra, and the hind wings for flying are compactly housed under the elytra. Unlike many other winged insects with exposed wings for flying, beetles can hide in various protected places without damaging their fragile hind wings. Thus, the structure of the beetle body may have particular merits, whereby the importance of wings for escaping from environmental harshness and natural enemies has been reduced in favor of protected bodies, and the cost of having fragile wings (i.e., avoiding damage) has been reduced by the innovative covering by the elytra. Overall, beetles may have been freed from the strong selection for maintaining the flight apparatus for survival compared to other winged insects.

The loss of flight ability occurs in various groups of beetles through degeneration of wing muscles and hind wings. Wingless species comprise more than 10 % of beetle species (Roff 1990), but the proportion of wingless species is higher in some groups such as Carabidae (Darlington 1943). In addition, flightless winged species exist that lack the flight muscles. Ikeda et al. (2012) hypothesized that flightless lineages are more genetically diverged among local populations due to restricted gene flow and hence have higher potential for allopatric speciation than lineages of flight-capable beetles. This was tested using the carrion beetle group Silphinae, which comprises both flight-capable and flightless species (Fig. 3.2).

We examined eight species of Japanese Silphinae, which consisted of five flight-capable species, two flightless (wingless) species, and one species with dimorphism in the flight muscles. Ecological correlates with flight ability and evolutionary and phylogeographic aspects of flight loss have been studied for these species (Ikeda et al. 2007, 2008, 2009; Ikeda and Sota 2011). First, we demonstrated that genetic differentiation among populations was much higher in flightless than flight-capable species using mitochondrial *COI–II* gene sequences (Table 3.1); flight-capable species exhibited only a small divergence in gene sequence across the distribution range. The same patterns were also revealed for two nuclear gene sequences, although the molecular variances of these genes were much smaller than those of the mitochondrial genes (Table 3.1). We then identified the number of demes (potential species) within each species, which can be treated as "species," using the

Fig. 3.2 Winged (*left, Necrophila brunnicollis*), wingless (*center, Silpha longicornis*), and winged species with dimorphism in flight muscles (*right, Necrophila japonica*) of Silphinae in Japan. Photo by Teiji Sota

Table 3.1 Genetic differentiation among populations of Japanese silphine species with different flight abilities (after Ikeda et al. 2012)

Flight ability[a]	Percent molecular variance among populations[b]			Number of demes[c]
Species	COI-II	PepCK	Wg	
Flight capable				
Dendroxena sexcarinata	5.6	−5.6	14	1 (1)
Oiceoptoma subrufum	6.5	−24.8	−20.8	1 (1)
Oiceoptoma nigropunctatum	3.4	10.5	18.6	1 (1)
Necrodes littoralis	−9.7	4.5	0	3 (1)
Necrophila brunnicollis	−1.7	0	0	1 (1)
Dimorphic				
Necrophila japonica	39.5***	29.6***	21.8***	1 (1)
Flightless				
Silpha longicornis	97.0***	86.3***	90.8***	11 (11)
Silpha perforata	70.1***	−7.9	52.8***	3 (2)

***$P < 0.001$

[a]Flight capable, with wings and flight muscles; dimorphic, with wings and with or without flight muscles; flightless, without wings and flight muscles
[b]Percent of molecular variances were obtained by AMOVA
[c]The numbers of demes (potential species) were obtained by the GMYC approach based on COI-II data. Only the results using single thresholds for species delimitation are presented.
In parentheses are numbers corrected for overestimation

generalized mixed Yule coalescent (GMYC) approach (Pons et al. 2006; Monaghan et al. 2009), which identifies demes based on a gene tree. As a result, the number of demes was larger for flightless species (nominal species) than flight-capable species. For the entire lineage of Silphinae with 106 nominal species, we estimated the speciation and extinction rates for flight-capable and flightless conditions, and found that the speciation rate for the flightless condition is twice that for the flight-capable condition. These results confirmed our prediction that allopatric differentiation and subsequent speciation are promoted by loss of flight.

3.2.2 Effects of Habitat Continuity and Persistence/Stability

Another factor affecting genetic differentiation among geographic populations is habitat discontinuity (e.g., Hastings and Harrison 1994; Brouat et al. 2003). If dispersal power is identical for two species, the one in a more discontinuous habitat will have a lower rate of gene flow and thus a higher probability of allopatric differentiation among geographic populations. Therefore, examining the effect of flight ability on geographic differentiation must include an index of habitat discontinuity. In addition, based on the generalization of dispersal strategy in relation to habitat persistence or stability (Southwood 1977), Ribera et al. (2001) predicted that genetic differentiation among local populations and the speciation rate will be higher for lineages living in more persistent habitats, for which selection against dispersal propensity results in lower gene flow among local populations (see also Papadopoulou et al. 2009). This habitat type hypothesis of speciation rate is difficult to test for the Silphinae because evaluating habitat stability for each species is complicated; silphine beetles live in various types of habitats ranging from unstable riverine sites to relatively stable forests, and a consistent difference in stability does not appear to exist among species' habitats.

Ikeda et al. (2012) took a different approach to assess the relationships between habitat discontinuity and persistence and the potential for speciation (i.e., genetic differentiation among populations). We conducted a generalized linear mode analysis to examine the additional effects of habitat discontinuity and habitat persistence in a geological timescale in terms of predicted geographic distribution areas both in the present and past (at the Last Glacial Maximum, LGM). The present and past distribution areas of each species were predicted based on climatic data, and the discontinuity of distribution areas was evaluated using an index of aggregation. As a result, the flight ability and the discontinuity of the present distribution areas were found to have a significant effect on the divergence of *COI–II* gene sequences, whereas the discontinuity of the past distribution areas had no effect. In addition, the number of GMYC-delimited demes was significantly affected by flight ability but not by the persistence of the distribution areas from the LGM to the present. From these results, we determined that flight ability is the primary factor determining the level of genetic differentiation among populations, although discontinuity of the present habitat also had a significant effect. Note that our evaluation of habitat discontinuity was indirect (i.e., based on a geographic information systems approach) and of a relatively large geographic scale (unit = 1-km mesh), and hence we likely have mainly detected the effects of large-scale dispersal by silphine beetles.

Although we have not studied the effect of fine-scale habitat continuity and stability, flight-capable and flightless silphine species live in the same habitats, and the evolutionary divergence of flight ability is based on their different diets (vertebrate carcasses for flight-capable, terrestrial invertebrates for flightless; Ikeda et al. 2008). Thus, we concluded that the flight loss hypothesis of the speciation rate is not directly related to the habitat type hypothesis, which implies the intimate relationship between habitat type and flight ability (Ikeda et al. 2012). Vogler and Timmermans (2012) argued that carcass feeding represents unstable

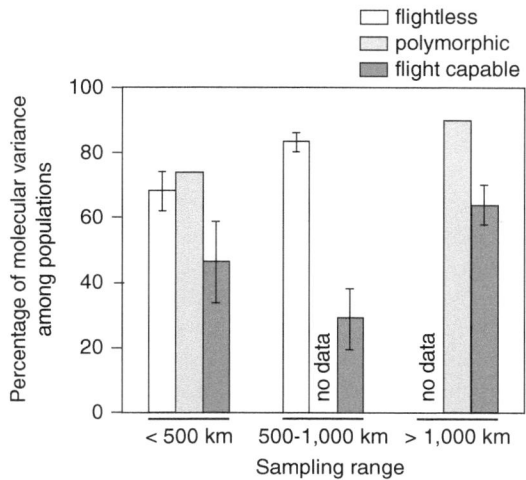

Fig. 3.3 Percentages of molecular variances among populations (±SE) of flightless, polymorphic, and flight-capable beetles in relation to sampling ranges. Modified from Ikeda et al. (2012)

habitat conditions, whereas soil animal feeding constitutes stable habitat conditions for the silphine beetles, and habitat stability could be the ultimate factor affecting the speciation rate in these beetles. They also introduced the example of flightless darkling beetles (Papadopoulou et al. 2009) in which species using stable habitat types exhibited higher genetic differentiation than those using unstable habitat types, which suggests an association among the stabilities of habitat, organisms, and speciation. In this case, however, the absence of flight-capable darkling beetles precludes the opportunity to distinguish the effects of habitat stability and flight ability (dispersal power) on genetic differentiation. Vogler and Timmermans (2012) commented that flight is not the only way of dispersal and dispersal power in other agents (e.g., walking) evolve in association with habitat stability. A link may exist between habitat stability and dispersal propensity in general, and the speciation rate would be more or less affected by habitat stability via its effect on the evolution of dispersal. However, the notion of habitat stability is variable among species with diverse life histories and resource requirements. Therefore, generalizing the relationship between habitat type and speciation is difficult.

3.2.3   Flight Capability and Speciation Rate in a Broader Context

Our hypothesis of a higher speciation rate with flight loss was examined for only one group of beetles and is far from sufficient. To compensate for the weakness of our evidence, we conducted a meta-analysis for the geographic differentiation patterns of various beetle groups (51 species from 15 families) based on a literature survey (Fig. 3.3; Ikeda et al. 2012). As for the population genetic analysis for Silphinae using *COI–II* genes, we examined the percent molecular variance of mitochondrial gene sequences (*COI–II* and its adjacent gene regions) among populations in a total molecular variance including within and among populations. In this analysis, we found that data for flight-capable species were collected from wider geographic areas than those for flightless species. This should reflect that flight-capable species had wider distribution ranges than flightless species. When adjusted for the distribution range size, the proportion of molecular variance among populations (Va) was larger for flightless than flight-capable species. In this case, Va tended to increase with increasing habitat range in flightless species, suggesting that range expansion per se increases the chance of allopatric differentiation.

The relationships among flight (dispersal) power, distribution range size, and speciation (and extinction) rate(s) should thus be compared among different lineages. Although some insect groups completely lack wings, the overall proportion of wingless species of insect is around 10 %, suggesting that the loss of wings does not provide an overwhelming advantage. The evolutionary dynamics of flight-capable and flightless lineages with respect to speciation and extinction rates requires more careful consideration. Flightless species may be vulnerable to unpredictable habitat disturbance and catastrophes compared to flight-capable species, and hence overall species richness in a flightless lineage may not be high despite a potentially high speciation rate if the extinction rate is high. However, vulnerability to disturbance and catastrophes at a regional scale, rather than very small local scales, may not differ much between flightless and flight-capable species. The stability of habitats likely greatly affects the survival of both flightless and flight-capable lineages, but generalizing the relationships among habitat stability, flight ability, and the extinction probability of species is very complicated.

Evolutionary loss of flight may decrease or eliminate the possibility of colonizing new regions that had not been previously occupied by the lineage. However, in the long term, the dispersal power of flightless lineages may not necessarily be low even across seemingly discontinuous topologies, such as those separated by large rivers or oceans. In fact, some flightless insects or weak flyers have reached even oceanic islands (e.g., by rafting, wind-borne dispersal) and achieved speciation. Thus, the dynamics of the number of species from lineages with and without flight abilities would be affected by multiple factors and needs to be examined more theoretically and empirically using enhanced methodologies for phylogeographic analyses.

3.3 Allochronic Speciation by Climatic Disruption of the Reproductive Period

The idea that climatic disruption of the reproductive period within a population will result in temporal reproductive isolation and eventually speciation was first invoked by Alexander and Bigelow (1960). They postulated that in a pair of field cricket species, disruptive selection for alternative cold-resistant overwintering stages resulted in segregation of the reproductive period, and divergence into two species with different seasonal cycles occurred in sympatry. Mayr (1963) argued that allochronic populations of the crickets could be formed by secondary contact of species, which were formed in different regions with different methods of selection favoring alternative life cycles. As a later phylogenetic study revealed that these crickets were not sister species (Huang et al. 2000), the divergence of these crickets may not have occurred as Alexander and Bigelow (1960) supposed. Nevertheless, their idea about speciation due to climatically induced temporal isolation is worthy of more detailed examination given the diversity of organisms living in seasonal environments.

Indeed, several recent studies have reported possible cases of allochronic speciation in insects (Filchak et al. 2000; Simon et al. 2000; Abbot and Withgot 2004; Santos et al. 2007, 2011) and other organisms (Friesen et al. 2007; Papa et al. 2007; Tomaiuolo et al. 2007; Devaux and Lande 2008). In some of these cases, however, the temporal isolation is caused by differential ecological adaptations, and the divergence of allochronic populations is a by-product of adaptation to resources with different seasonality. A well-known example is sympatric speciation in the apple maggot *Rhagoletis pomonella*, for which adaptation to different hosts (hawthorn and apple) is associated with differences in mating site and time (Filchak et al. 2000). Speciation due to by-product temporal isolation is not considered allochronic speciation. No clear case of climatic disruption of allochronic populations has been observed since Alexander and Bigelow (1960) first proposed the mechanism, and whether/how often temporal isolation can occur without ecological adaptation to different resources has been questioned.

3.3.1 Incipient Allochronic Speciation in a Winter Moth

A winter geometric moth *Inurois punctigera* (Fig. 3.4) in Japan is known to have two separate flight periods in early and late winter at some localities (Nakajima 1998). Because adult moths live for only 2 weeks, the moths occurring in early and late winter obviously represent different cohorts. Satoshi Yamamoto, who started as a graduate student at Kyoto University in 2005, was interested in this phenomenon and hypothesized that *I. punctigera* populations are being disrupted by midwinter harshness and undergoing allochronic speciation. Since then, he has conducted extensive field sampling and phylogenetic analysis to test his hypothesis.

I. punctigera occurs throughout the main islands of Japan, from Hokkaido to Kyushu. The flight season shifts from a typical disrupted pattern (early and late winter) to a continuous pattern (single period) with increasing average temperature during the winter (Fig. 3.5). Localities with the disrupted flight periods have been found throughout Japan, not only at high latitudes but also at high elevations at relatively lower latitudes. *Inurois* larvae are generalist feeders on deciduous broadleaf trees. Larvae of early and late winter moths occur simultaneously in the spring and pupate by early summer. They differ in the length of pupal period and emerge as adults either in early or late winter. Adults do not eat, and no difference in ecological requirements appears to exist between early and late winter moths. Thus, the life

Fig. 3.4 Copulating female (*upper*) and male (*lower*) winter geometrid moth *Inurois punctigera*. Photo by Satoshi Yamamoto

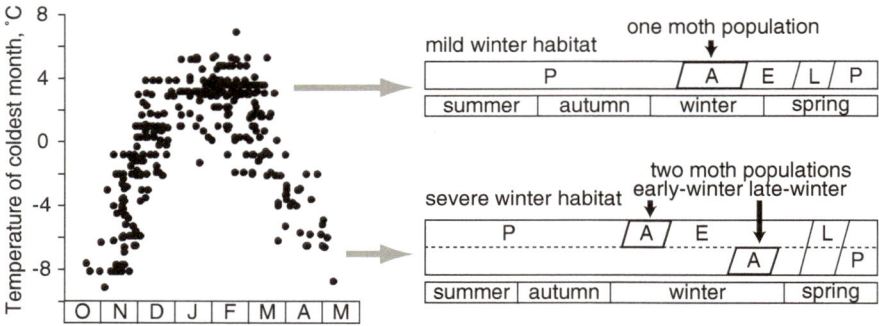

Fig. 3.5 Emergence patterns and life cycles of *Inurois punctigera*. On the *left*, the dates of the appearance of males are plotted against the mean temperature of the coldest month at the locality. Corresponding life cycles for mild winter (only one population) and severe-winter habitats (two allochronic populations) are shown on the *right*. *A* adults, *E* eggs, *L* larvae, *P* pupae. Modified from Yamamoto and Sota (2012)

cycles in *I. punctigera* are not related to adaptations to seasonally different resources (Yamamoto and Sota 2009).

Phylogenetic analyses using mitochondrial gene sequences revealed that the early and late winter moths have diverged genetically (Yamamoto and Sota 2009). A closer examination of four localities with both types of moths in Honshu using mitochondrial gene sequence and amplified fragment length polymorphism (AFLP) markers revealed that populations of each temporal type were closely related to one

Fig. 3.6 Genetic differentiation (F_{st}) among weekly cohorts of *Inurois punctigera* at Kyoto, showing isolation-by-time patterns. (**a**) AFLP data. (**b**) Mitochondrial COI data. Mantel test showed that the correlation between F_{st} and time lag was significant for both types (AFLP, $r=0.286$, $P=0.033$; *COI*, $r=0.577$, $P=0.001$)

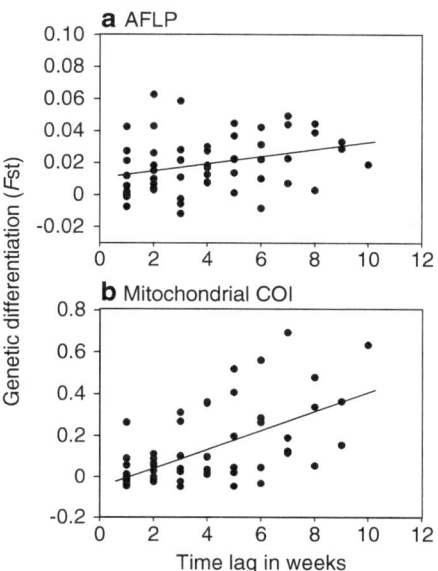

another, but sympatric populations of the two temporal types were not closely related to each other (Yamamoto and Sota 2009). This indicates that the allochronic lineages have originated with limited opportunities in the past and have spread to various places.

Additional important insights have come to light by examining the genetic differentiation between temporal cohorts in a habitat with a relatively mild winter and an extended emergence period (Yamamoto and Sota 2009). In Kyoto, the flight period extends from December through March, and moths would be replaced gradually because each adult lives for 2 weeks at most. Examination of the genetic divergence between weekly cohorts revealed the "isolation-by-time" pattern (Hendry and Day 2005) in both mitochondrial and AFLP markers (Fig. 3.6). This finding suggests that the Kyoto population maintained some genetic variation regarding the timing of emergence, and that genetic variation is based on some additive factors. The present Kyoto population may have formed through a secondary contact between early and late winter moths due to global warming, as the mean January temperature has increased by about 2 °C during the past 100 years and the climatic conditions of Kyoto are not much different from that of Gifu with allochronic moth populations (Yamamoto and Sota, unpublished). Therefore, the genetic variation in the timing of emergence might have resulted from the hybridization of early and late winter genotypes.

Because changes in climatic conditions, unlike changes in biological conditions, would affect organisms in different regions in a similar way, the development of allochronic winter moths within *I. punctigera* could occur repeatedly in different regions of Japan. This has been found in a subsequent survey (Yamamoto and Sota 2012).

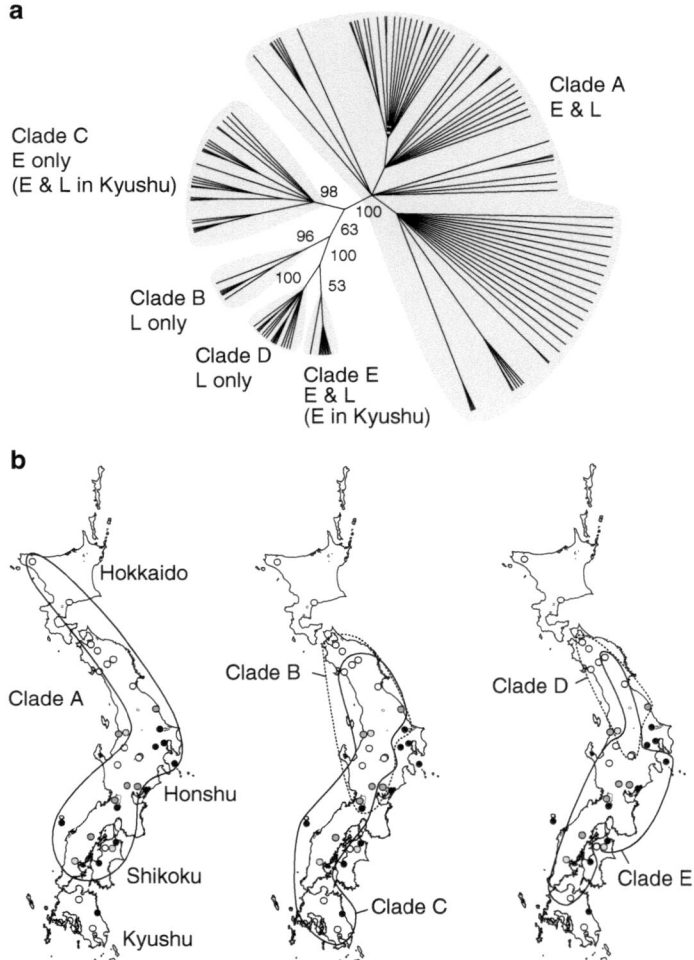

Fig. 3.7 Phylogeny of *COI* haplotypes of *Inurois punctigera* with five clades (**a**) and the geographic distribution of clades (**b**). In (**a**), E refers to the early winter moth and L to the late winter moth. Modified from Yamamoto and Sota (2012). Only C and E haplotypes were found in Kyushu. C haplotypes were possessed only by late-winter moths in Honshu but by both early and late winter moths in Kyushu, suggesting a shift from early to late winter moths in Kyushu

We focused on the divergence of early and late winter moths in Kyushu, which could be an occurrence of allochronic divergence independent of that in Honshu (Fig. 3.7). In mitochondrial *COI* haplotypes, *I. punctigera* in Honshu exhibited five haplotype clades (A–E), all of which were present in late winter moths, whereas only A and E were present in early winter moths. In Kyushu, only clade C and E haplotypes were found, and most moths (both early and late winter) possessed haplotypes of C clade, which was exclusively the clade of late winter moths in Honshu. Therefore, early winter moths in Kyushu are considered to have evolved independently from late winter moths with a clade C haplotype. In situ divergence in Kyushu was also indicated by AFLP data, which showed a sister relationship of the early and late winter moth populations in Kyushu.

3.3.2 An Extended View of Allochronic Speciation

Speciation by temporal isolation has almost exclusively been discussed within a framework of sympatric speciation (Alexander and Bigelow 1960), and evidence for the occurrence of allochronic speciation has been sought for sympatric pairs of allochronic populations. Nevertheless, temporal barriers causing reproductive isolation between individuals of a species do not preclude the formation of allopatric and temporally isolated adjacent groups of individuals. This can be the case for small insects with limited dispersal abilities and the action of a climatic barrier, such as a cold winter or hot and dry summer that affects activity. During a climatic change with the increasing strength of such a temporal barrier, the reproductive period of an insect population may shift to before or after this barrier, or may be disrupted into both. However, how the "before" and "after" groups (populations) establish may be subject to demographic stochasticity or may depend on the genetic constitution of the life cycle timing of the original populations. As a result, in an area with similar climatic conditions, one location may harbor both the "before" and "after" individuals, but other areas may harbor either the "before" or "after" individuals (Fig. 3.8). In addition, movement of individuals between locations will cause a temporal change in life cycles at each location. Thus, allochronic speciation, if it occurs, is not restricted to sympatric species. In a broad sense, the divergence of allochronic populations in a region by a common temporal barrier (such as winter cold) can be classified as allochronic speciation irrespective of strict sympatry.

3.3.3 Potential Importance of Temporal Isolation By Climatic Disruption in the Divergence of Insects

Our study of I. punctigera suggested the potential importance of temporal isolation by harsh winter conditions as a speciation mechanism in other winter moths. The genus Inurois occurs in Japan and the East Asian mainland, and consists of both early and late winter species (Beljaev 1996), and the divergence of early and late winter species seems to have occurred repeatedly (Yamamoto and Sota 2009). Moths with a winter emergence appear in several different subfamilies within Geometridae and other families of moths (Yamamoto and Sota 2007). Therefore, similar cases of disruptive divergence by winter harshness may exist in different group of winter moths.

Fig. 3.8 Divergence pattern of local sites harboring populations after a climatic disruption event. Local sites consist of those with two allochronic populations or populations with either temporal type

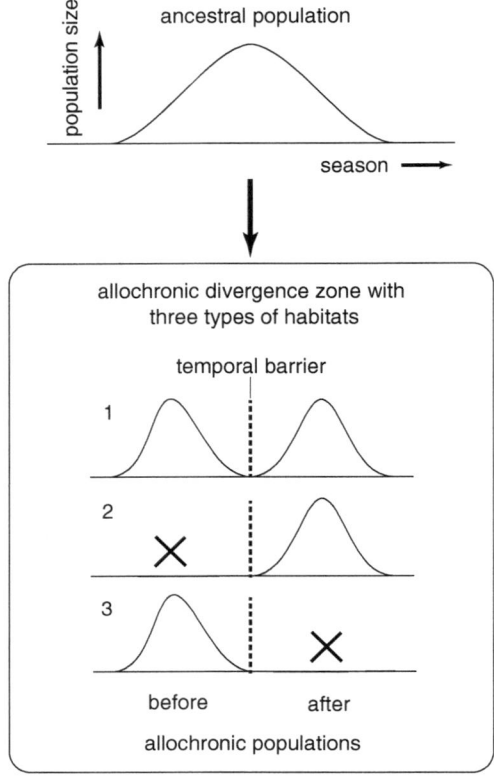

Univoltine insects in seasonal environments tend to diverge in reproductive season either in spring or autumn, avoiding harsh climatic conditions such as during the cold of winter and hot and dry midsummer (Tauber et al. 1986). Divergence in the timing of reproduction before and after the climatic harshness will cause allochronic speciation. However, the change in reproductive timing between spring and autumn with a definitely unfavorable midwinter would be difficult because this change requires different kinds of adaptation (e.g., in diapause, resource use) and is unlikely to occur frequently in most groups of insects. The repeated occurrence of allochronic divergence in *I. punctigera* was possible because this moth was originally adapted to reproduction in winter and required changes in life history were minimal, requiring only changes in the timing of adult emergence. For most insects, the divergence of spring/autumn reproduction by winter disruption may have occurred rarely in association with the colonization of different climatic zones or global climatic changes. Therefore, analyzing the phylogenetic pattern of divergent reproductive timing in various insect groups is important to understand how such life history changes have occurred and contributed to species diversification.

3.4 Conclusions

Our recent studies on speciation processes in space and time have been introduced. The high species diversity on our planet is undoubtedly caused by the high heterogeneity and discontinuity of environments in space and time and the variety of life history tactics that each organism can make to adapt. The diversity of strategies by which organisms cope with environmental heterogeneity spurs the diversification of organisms. Revealing the various mechanisms of diversification is an important task for biodiversity studies and will be useful for maintaining the diversity of living organisms on our planet.

Acknowledgements I thank Hiroshi Ikeda and Satoshi Yamamoto for their collaboration and reviewing of this manuscript. Our studies of beetles and winter moths were supported by the Global COE Program A06, "Formation of Strategic Base for Biodiversity and Evolutionary Research: from Genomics to Ecosystems" from the Ministry of Education, Culture, Sports, Science and Technology of Japan. We thank especially K. Agata, the project leader, and E. Kawaguchi, the sequencing manager, for their continuous support.

References

Abbot P, Withgot JH (2004) Phylogenetic and molecular evidence for allochronic speciation in gall-forming aphids (*Pemphigus*). Evolution 58:539–553

Alexander RD, Bigelow RS (1960) Allochronic speciation in field crickets, and a new species, *Acheta veletis*. Evolution 14:334–346

Beljaev EA (1996) "Winter" geometridae (Lepidoptera) of Japan sea region: taxonomic composition, morphological and biological features, biogeographic analysis. Chtena Pamyati Alekseya Ivanovicha Kurentsova 6:33–76

Brouat C, Sennedot F, Audiot P, Leblois R, Rasplus J-Y (2003) Fine-scale genetic structure of two carabid species with contrasted levels of habitat specialization. Mol Ecol 12:1731–1745

Coyne JA, Orr HA (2004) Speciation. Sinauer Associates, Sunderland

Darlington PJ Jr (1943) Carabidae of mountains and islands: data on the evolution of isolated faunas, and on atrophy of wings. Ecol Monogr 13:37–61

Devaux C, Lande R (2008) Incipient allochronic speciation due to non-selective assortative mating by flowering time, mutation and genetic drift. Proc R Soc B 275:2723–2732

Dieckmann U, Doebeli A, Metz JAJ, Tautz D (eds) (2004) Adaptive speciation. Cambridge University Press, Cambridge

Engel MS, Grimaldi DA (2004) New light shed on the oldest insect. Nature 427:627–630

Farrell BD (1998) "Inordinate fondness" explained: why are there so many beetles? Science 281:555–559

Filchak KE, Roethele JB, Feder JF (2000) Natural selection and sympatric divergence in the apple maggot *Rhagoletis pomonella*. Nature 407:739–742

Friesen VL, Smith AL, Gómez-Diaz E, Bolton M, Furness RW, González-Solís J, Monteiro LR (2007) Sympatric speciation by allochrony in a seabird. Proc Natl Acad Sci USA 104: 18589–18594

Grimaldi D, Engel MS (2005) Evolution of the insects. Cambridge University Press, Cambridge

Hansen TA (1983) Modes of larval development and rates of speciation in early tertiary neogastropods Science 220:501–502

Hastings A, Harrison S (1994) Metapopulation dynamics and genetics. Annu Rev Ecol Syst 25:167–188

Hendry AP, Day T (2005) Population structure attributable to reproductive time: isolation by time and adaptation by time. Mol Ecol 14:901–916

Huang Y, Ortí G, Sutherlin M, Duhachek A, Zera A (2000) Phylogenetic relationships of North American field crickets inferred from mitochondrial DNA data. Mol Phylogenet Evol 17:48–57

Hunt T, Bergsten J, Levkanicova Z, Papadopoulou A, John O, St WR, Hammond PM, Ahrens D, Balke M, Caterino MS, Gomez-Zurita J, Ribera I, Barraclough TG, Bocakova M, Bocak L, Vogler AP (2007) A comprehensive phylogeny of beetles reveals the evolutionary origins of a superradiation. Science 318:1913–1916

Ikeda H, Sota T (2011) Macro-scale evolutionary patterns of flight muscle dimorphism in the carrion beetle *Necrophila japonica*. Ecol Evol 1:97–105

Ikeda H, Kubota K, Kagaya T, Abe T (2007) Flight capability and food habits of silphine beetles: are flightless species really "carrion beetles"? Ecol Res 22:237–241

Ikeda H, Kagaya T, Kubota K, Abe T (2008) Evolutionary relationships among food habit, loss of flight, and reproductive traits: life history evolution in the Silphinae (Coleoptera: Silphidae). Evolution 62:2065–2079

Ikeda H, Kubota K, Cho YB, Liang H, Sota T (2009) Different phylogeographic patterns in two Japanese Silpha species (Coleoptera: Silphidae) affected by climatic gradients and topography. Biol J Linn Soc 98:452–467

Ikeda H, Nishikawa M, Sota T (2012) Loss of flight promotes beetle diversification. Nat Comm 3:648

Jablonski D (1986) Larval ecology and macroevolution in marine invertebrates. Bull Mar Sci 39:565–587

Lövei GL, Sunderland KD (1996) Ecology and behavior of ground beetles (Coleoptera: Carabidae). Annu Rev Entomol 41:231–256

Mayr E (1963) Animal species and evolution. Harvard University Press, Cambridge

Monaghan MT, Wild R, Elliot M, Fujisawa T, Balke M, Inward DJG, Lees DC, Ranaivosold R, Eggleton P, Barraclough TG, Vogler AP (2009) Accelerated species inventory on Madagascar using coalescent-based models of species delineation. Syst Biol 58:298–311

Nakajima H (1998) A taxonomical and ecological study of the winter geometrid moths (Lepidoptera, Geometridae) from Japan. Tinea 15(suppl 2):1–246

Papa R, Israel JA, Marzano FN, May B (2007) Assessment of genetic variation between reproductive ecotypes of Klamath River steelhead reveals differentiation associated with different run-timings. J Appl Ichtyol 23:142–146

Papadopoulou A, Anastasiou I, Keskin B, Vogler AP (2009) Comparative phylogeography of tenebrionid beetles in the Aegean archipelago: the effect of dispersal ability and habitat preference. Mol Ecol 18:2503–2517

Pons J, Barraclough TG, Gomez-Zurita J, Cardoso A, Duran DP, Hazell S, Kamoun S, Sumlin WD, Vogler AP (2006) Sequence-based species delimitation for the DNA taxonomy of undescribed insects. Syst Biol 55:595–609

Ribera I, Barraclough TG, Vogler AP (2001) The effect of habitat type on speciation rates and range movements in aquatic beetles: inferences from species-level phylogenies. Mol Ecol 10:721–735

Roff DA (1986) The evolution of wing dimorphism in insects. Evolution 40:1009–1020

Roff DA (1990) The evolution of flightlessness in insects. Ecol Monogr 60:389–421

Roff DA (1991) Life history consequences of bioenergetic and biomechanical constraints on migration. Am Zool 31:205–215

Roff DA (1994) The evolution of flightlessness: is history important? Evol Ecol 8:639–657

Roff DA, Fairbairn DJ (1991) Wing dimorphisms and the evolution of migratory polymorphisms among the Insecta. Am Zool 31:243–251

Rundle HD, Nosil P (2005) Ecological speciation. Ecol Lett 8:336–352

Santos H, Rousselet J, Magnoux E, Paiva MR, Branco M, Kerdelhué C (2007) Genetic isolation through time: allochronic differentiation of a phenologically atypical population of the pine processionary moth. Proc R Soc B 274:935–941

Santos H, Burban C, Rousselet J, Rossi JP, Branco M, Kerdelhué C (2011) Incipient allochronic speciation in the pine processionary moth (*Thaumetopoea pityocampa*, Lepidoptera, Notodontidae). J Evol Biol 24:146–158

Simon C, Tang J, Dalwadi S, Staley G, Deniega J, Thomas RU (2000) Genetic evidence for assortative mating between 13-year cicadas and sympatric "17-year cicadas with 13-year life cycles" provides support for allochronic speciation. Evolution 54:1326–1336

Smith CI, Farrell BD (2006) Evolutionary consequences of dispersal ability in cactus-feeding insects. Genetica 126:323–334

Sobel JM, Chen G, Watt LR, Schemske DW (2009) The biology of speciation. Evolution 64:295–315

Southwood TRE (1977) Habitat, the templet fro ecological strategies? J Anim Ecol 46:337–365

Tauber MJ, Tauber CA, Masaki S (1986) Seasonal adaptations of insects. Oxford University Press, New York

Tomaiuolo M, Hansen TF, Levitan DR (2007) A theoretical investigation of sympatric evolution of temporal reproductive isolation as illustrated by marine broadcast spawners. Evolution 61:2584–2595

Vogler AP, Timmermans MJTN (2012) Speciation: don't fly and diversity? Curr Biol 22:R284

Wagner DL, Liebherr JK (1992) Flightlessness in insects. Trends Ecol Evol 7:216–220

Yamamoto S, Sota T (2007) Phylogeny of the Geometridae and the evolution of winter moths inferred from a simultaneous analysis of mitochondrial and nuclear genes. Mol Phyol Evol 44:711–723

Yamamoto S, Sota T (2009) Incipient allochronic speciation by climatic disruption of the reproductive period. Proc R Soc B 276:2711–2719

Yamamoto S, Sota T (2012) Parallel allochronic divergence in a winter moth due to disruption of reproductive period by winter harshness. Mol Ecol 21:174–183

Zera AJ (1981) Genetic structure of two species of waterstriders (Gerridae: Hemiptera) with differing degrees of winglessness. Evolution 35:218–225

Zera AJ, Denno RF (1997) Physiology and ecology of dispersal polymorphism in insects. Annu Rev Entomol 42:207–231

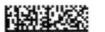